Hans-Jörg Bart

Reactive Extraction

With 101 Figures

Springer

Series Editors

Prof. Dr.-Ing. Dieter Mewes
Universität Hannover
Institut für Verfahrenstechnik
Callinstraße 36
30167 Hannover, Germany

Prof. Dr.-Ing. E. h. Franz Mayinger
Technische Universität München
Lehrstuhl A für Thermodynamik
Boltzmannstraße 15
85748 Garching, Germany

Author

Prof. Dipl.-Ing. Dr. Hans-Jörg Bart
Universität Kaiserslautern
FB Maschinenbau und Verfahrenstechnik
Lehrstuhl für Thermische Verfahrenstechnik
Gottlieb-Daimler-Straße
67663 Kaiserslautern, Germany

ISBN 3-540-41087-2 Springer-Verlag Berlin Heidelberg New York

Library of Congress Cataloging-in-Publication Data applied for
Die deutsche Bibliothek – CIP-Einheitsaufnahme

Bart, Hans-Jörg:
Reactive extraction / Hans-Jörg Bart. – Berlin ; Heidelberg ; New York ; Barcelona ; Hong Kong ; London; Milan ; Paris ; Singapore ; Tokyo : Springer, 2001
(Heat and mass transfer)
(Engineering online library)
ISBN 3-540-41087-2

Springer-Verlag Berlin Heidelberg New York
a member of BertelsmannSpringer Science+Business Media GmbH

Printed in Germany

Typesetting: Camera-ready by author
Cover design: de'blik, Berlin

Printed on acid-free paper SPIN: 10775055 61/3020 CU – 5 4 3 2 1 0 –

Preface

This booklet is designed to bridge the gap between handbooks and technical literature and aims at graduate students or experienced readers. Commercial flowsheeting simulation software is increasingly available and is used in the early steps of process design in industry. As to this, more sophisticated and precise models based on activities instead of concentrations should be used. After an introductory chapter there is in Chapter 2 an intensive discussion of reactive phase equilibria of ionic and non-ionic solutes based on chemical potentials. Chapter 3 introduces to multicomponent diffusion and mass transfer. However, the main focus is on the reactive mass transfer on rigid and mobile surfaces where the interfacial reaction, molecular diffusion and adsorption layers are decisive. The respective extraction of zinc with a cation exchanger and of acetic acid with an anion exchanger is discussed as case studies. Since adsorption layers and surfactants have a major impact on liquid–liquid extraction efficiency, the final chapter reviews several techniques which make use of polymeric species in an extractive process. A short review is also given on extraction apparatus and the hydrodynamics (hydraulic design, droplet populance balances) of columns.

Much of the booklet is based on the PhD works of C. Czapla (2000), G. Modes (2000), H. Klocker (1996), T. Kronberger (1995), M. Mörters (2000), M. Roos (2000), M. Traving (2000) and B. Wachter (1996) who I wish to thank for their fruitful contributions. I also appreciate the stimulating discussions I have had with Dr. M.J. Slater (University of Bradford, Great Britain) and Prof. J.A. Wesselingh (University of Groningen, The Netherlands) which helped clarify my thinking and led me to new ways of presenting some of the material. Special thanks go to my wife Ariane and my secretary I. Behrendt for typing and retyping the drafts, to Prof. Dr. J. Draxler (Montanuniversität Leoben, Austria) for proof-reading the text and to my co-worker D. Bosse who helped with the text, figures and final tedious layout of the book.

Kaiserslautern, September 2000 **Hans-Jörg Bart**

Table of Contents

Nomenclature

A	Interfacial area [m^2]
A, B, C	Margules parameter [-]
A_d	Area downcomer [m^2]
A_h	Area of holes [m^2]
A_T	Tray area [m^2]
A_m	Debye–Hückel constant [m$^{3/2}$mol$^{-1/2}$]
a	Specific interfacial area [m^2/m^3]
a	Ionic radius [m]
a_i	Activity of species i [-]
a_w	Activity of water [-]
$a_1, a_2, ..$	Coefficient [-]
B_{ii}, B_{ij}	Function of the Maxwell–Stefan diffusion coefficients and mol fractions of a mixture [s/m^2]
C_{IP}	Instability constant [-]
C_k	Constant [depends]
$C_1, \ldots, C_4$	Kinetic constants [depends]
C_{ii}, C_{ij}	Density of energy of cohension [cal/cm^3]
C^{ϕ}	Pitzer parameter [-]
$C_0, \ldots, C_4$	Solvatochromic parameter [-]
c_i	Molar density of species i [mol/L]
c_s	Molar density of solvent [mol/L]
c_t	Molar mixture density [mol/L]
$c_{i,0}$	Initial molar density [mol/L]
$c^{\oplus}$	Standard molarity (1 mol/L) [mol/L]
D	Distribution coefficient [-]
D	Diameter [m]
D	Fick's diffusion coefficient (binary) [m^2/s]
$D2EHPA$	Di(2-ethylhexyl) phosphoric acid
D_{ij}	Fick's diffusion coefficient[m^2/s]

D_{ij}^{0}	Diffusion coefficient at infinite dilution ($x_i \rightarrow 0$) [m^2/s]
D_{ax}	Axial dispersion coefficient [m^2/s]
D_k	Column diameter [m]
$Đ$	Maxwell–Stefan diffusion coefficients (binary) [m^2/s]
$Đ_{ij}$	Maxwell–Stefan diffusion coefficients [m^2/s]
D_R	Rotor diameter [m]
D_S	Stator diameter [m]
d	Droplet diameter [m^{-1}]
d_i	Driving force for mass diffusion [m^{-1}]
d_S	Sauter diameter [m]
d_h	Hole diameter [m]
E	Activation energy [J/mol]
E	Eddy diffusivity [m^2s^{-1}]
e	Elementary charge (1.602×10^{-19} C) [C]
F_i	Force per mol of species *i* [N/mol]
F_i	Force per kg [N/kg]
F_R	Friction force [m^{-1}]
f	Fugacity [Pa]
f_{12}	Friction coefficient [sm^{-2}]
$\mathfrak{I}$	Faraday constant ($9.65 \cdot 10^4 \times$ C/mol) [C/mol]
G	Gibbs energy [J]
G_{ij}	NRTL Boltzmann factor [-]
G^E	Gibbs excess energy [J]
G^{ID}	Gibbs energy (ideal mixture) [J]
g	Partial molar Gibbs energy [J/mol]
g	Acceleration due to gravity (9.81 m/s^2) [m/s^2]
$g(d)$	Break-up frequency of a droplet with diameter *d* [s^{-1}]
g^c	UNIQUAC Gibbs combinatorial energy [J/mol]
g^E	Partial molar Gibbs excess energy [J/mol]
g^{ID}	Partial molar Gibbs energy (ideal mixture) [J/mol]
g_{ij}	NRTL Gibbs residual energy [J/mol]
g^R	UNIQUAC Gibbs residual energy [J/mol]
H	Enthalpy [J]
HAc	Acetic acid
HCl	Hydrochloric acid

H_c	Compartment height [m]
H^{ID}	Enthalpy (ideal mixture) [J]
H^E	Excess enthalpy [J]
H_i	Henry constant [Pa]
H_K	Column height [m]
H_T	Tray spacing [m]
h	Partial molar enthalpy [J/mol^{-1}]
h_c	Height of coalescence layer [m]
h^{ID}	Partial molar enthalpy (ideal mixture) [J/mol^{-1}]
h^E	Partial molar excess enthalpy [J/mol^{-1}]
I	Ionic strength [mol/kg]
I	Current density of a mixture [A/m^2]
J_i	Molar diffusion flux [mol m^{-2} s^{-1}]
J_i^n	Molar diffusion flux relative to the solvent velocity [mol m^{-2} s^{-1}]
K	Equilibrium constant [depends]
k	Mass transfer coefficient [m/s]
$k^\bullet$	High flux mass transfer coefficient [m/s]
k	Boltzmann constant (1.38066×10^{-23} J K^{-1}) [J K^{-1}]
k_v	Kinetic constant (forward reaction) [depends]
$k_v^\bullet$	Pseudo kinetic constant (forward reaction) [depends]
k_r	Kinetic constant (reward reaction) [depends]
$k_r^\bullet$	Pseudo kinetic constant (reward reaction) [depends]
k_f	Friction factor [-]
L_c	Characteristic length [m]
l_{ij}	Binary parameter [-]
l	Film thickness [m]
l_o	Film thickness (org. phase) [m]
l_w	Film thickness (aqu. phase) [m]
MIBK	Methyl isobutyl ketone
M_i	Molar weight of species *i* [g/mol]
M_w	Molar weight of water [g/mol]
m	Contamination factor [-]
m_i	Molal density of species *i* [mol/kg]
$m^\oplus$	Standard molality (1 mol/kg) [mol/kg]

N	Revolutions per second [s^{-1}]
$NaDdS$	Sodium dodecane sulfonate
$NaLS$	Sodium lauryl sulfate
N_A	Avogadro number (6.02205×10^{23} mol^{-1}) [mol^{-1}]
N_E	Number of experiments [-]
N_i	Molar flux of a component i relative to a stationary reference frame [mol m^{-2} s^{-1}]
N_p	Number of model parameters [-]
N_t	Mixture molar flux relative to a stationary reference frame [mol m^{-2} s^{-1}]
N_h	Number of holes [-]
n	Number of species in a mixture [-]
n_I	Number of ionic species in a mixture [-]
n_i	Moles of species i [mol]
n_w	Moles of water [mol]
p	Pressure [Pa]
p_i	Partial pressure of component i [Pa]
p_i^o	Vapor pressure of component i [Pa]
p^L	Vapor pressure of a liquid mixture [Pa]
p^*	Standard pressure (0.1 MPa) [Pa]
P	Probability [-]
Q	Throughput [$m^3 s^{-1}$]
Q	Factor ($= p^p\ q^q$) [-]
Q_c	Continuous phase flow [$m^3 s^{-1}$]
Q_d	Dispersed phase flow [$m^3 s^{-1}$]
q_i	Van der Waals' surface [m^2]
r_i	Van der Waals radius [m]
R^0	Initial reaction rate [mol m^{-3} s^{-1}]
R	Gas constant (8.31441 J K^{-1} mol^{-1}) [J K^{-1} mol^{-1}]
$[RH]$	Ion exchanger concentration [mol/m^3]
$[R_2H_2]$	Ion exchanger dimer concentration [mol/m^3]
r	Radius [m]
S	Entropy [J K^{-1}]
S	Constant in the Masson equation $cm^3 mol^{-1} (mol/L)^{-1/2}$]

S_B	Source term breakage [-]
S_C	Source term coalescence [-]
s	Partial molar entropy [$Jmol^{-1}K^{-1}$]
s	Surface renewal frequency [s^{-1}]
T	Temperature [K]
TBP	Tributylphosphate
TOACl	Dodecyltrimethylammonium chloride
TOA	Trioctyl amine
TX100	Octylpoly(ethylglycolether)$_{10}$, Triton X-100
t	Time [s]
t_e	Exposure time [s]
t_j	Transference number [-]
U_h	Hole velocity [m/s]
U_c	Downcomer velocity [m/s]
u	Molar average reference velocity [m/s]
u_i	Velocity of diffusion of species *i* [m/s]
u_{ij}	UNIQUAC interaction parameter [J mol^{-1}]
u_n	Velocity of diffusion of solvent [m/s]
V	Volume [m^3]
V_i	Partial molar volume of component *i* [m^3/mol]
V_S	Slip velocity [m/s]
V_s	Partial molar volume of solvent [m^3/mol]
V_m	Partial mean molar volume [m^3/mol]
V_w	Partial molar volume of water [m^3/mol]
w	Velocity [m/s]
w_d	Droplet rising velocity [m/s]
w_T	Terminal velocity [m/s]
x	*X*-direction [m]
x_i	Mol fraction of component *i* [-]
Δx_i	Mol fraction difference between bulk and interface [-]
Y_i	Experimental value [depends]
$Y_{i,calc}$	Simulated value [depends]
z	*Z*-direction [m]
z_i	Charge of an ionic species [-]
z_d	Downcomer height [m]

Greek letters

α_i	Solvatochromic parameter [-]
α	Pitzer parameter [-]
α_{ij}	NRTL parameter [-]
α_L	Langmuir constant [mol/m^2]
$\beta(d_0,d)$	Daughter droplet distribution density [m^{-1}]
β_i	Solvatochromic parameter [-]
$\beta^{(0)}$, $\beta^{(1)}$	Binary Pitzer parameter [-]
Γ	Thermodynamic factor (binary) [-]
Γ_{ij}	Thermodynamic factor [-]
γ	Interfacial coverage [mol/m^2]
γ_i	Activity coefficient, mole fraction, sym. [-]
γ_i^H	Activity coefficient, mole fraction, non-sym. [-]
γ_i^c	Activity coefficient, molar scale, non-sym. [-]
γ^m	Activity coefficient, molal scale, non-sym. [-]
$\gamma^{\pm}$	Mean electrolyte activity coefficient [-]
γ^{∞}	Activity coefficient, infinite dilution [-]
γ_L	Langmuir constant [mol/m^3]
∂_i	Solvatochromic parameter [-]
δ_i	Hildebrand–Scott solubility parameter [cal$^{1/2}$cm$^{-3/2}$]
δ_{ij}	Kronecker delta ($\delta_{ij} = 1$, if $i = j$; $\delta_{ij} = 0$, if $i \neq j$) [-]
ε	Porosity [-]
ε_0	Dielectricity constant, vacuum (8.85419 J^{-1}C^{2}m^{-1}) [J^{-1}C^{2}m^{-1}]
ε_r	Dielectricity constant, relative to vacuum($\varepsilon = \varepsilon_0\, \varepsilon_r$) [-]
Φ_i	UNIQUAC average segment fraction [-]
Φ^V	Apparent molar volume of an ionic species [cm^3/mol]
Φ^0	Apparent molar volume of an ionic species at infinite dilution [cm^3/mol]
ϕ	Electrical potential [V]
ϕ	Hold-up [-]
φ_i	Fugacity coefficient [-]
ϕ_w	Osmotic coefficient [-]

η	Viscosity, dynamic [mPa s]
Θ_m	UNIFAC surface fraction [-]
Ψ	Electrical potential [V]
Ψ	Weisz–Prater criterion [-]
Ψ_0	Interfacial electrical potential [V]
Ψ_{nm}	UNIFAC parameter [-]
κ	Equivalent conductivity, mixture [S/m]
κ	Debye–Hückel constant [m^{-1}]
κ_j	Equivalent conductivity, species [S/m]
κ_v	Overall forward reaction rate constant [depends]
κ_r	Overall reward reaction rate constant [depends]
λ_i	Solvatochromic parameter [-]
λ_{ij}	Binary Pitzer parameter [-]
μ	Chemical potential [J/mol]
$\mu^o{}_i$, $\mu^\infty{}_i$, $\mu^\oplus{}_i$, μ^o	Standard chemical potential [J/mol]
$\Delta\mu_{s,i}^\infty$	Standard molar chemical potential of solvation [J/mol]
ν	Viscosity, kinematic [$m^2 s$]
ν_i	Stoichiometric coefficient [-]
$\nu_i^{(k)}$	Number of structural groups [-]
π^*	Solvatochromic parameter [-]
θ_i	UNIQUAC area fraction [-]
ρ_i	Density, species [kg/m^3]
ρ_L	Density, mixture [kg/m^3]
ρ_s	Density, solvent [kg/m^3]
ρ_w	Density, water [kg/m^3]
σ	Regression parameter [-]
σ	Interfacial tension [N/m]
τ_i	Solvatochromic parameter [-]
τ_{ij}	NRTL resp. UNIQUAC parameter [-]
τ_{ijk}	Ternary Pitzer parameter [-]
ω	Pitzer parameter [-]
ξ	Zeta potential [V]

Further symbols

( )	Vector
[ ]	Concentration [mol/L]
[ ]	Matrix
[I]	Identity matrix [-]
—	Overbar denotes organic species
$\nabla_{P,T}$	Gradient at constant pressure and temperature
Δ	Delta

Subscripts

ad	Interfacially adsorbed
aq	Aqueous
av	Average between bulk and interface
c	Continuous phase
crit	Critical
d	Dispersed phase
D	Downcomer
DIL	Diluent
eff	Effective
h	Hole
I	Interfacial
o	Initial
p	Particle
T	Terminal
T	Tray
tot	Total
w	Water

Superscripts

c	Referring to molarities
C	Combinatorial
E	Excess
G	Referring to gas phase
H	Referring to Henry law
ID	Ideal
L	Referring to liquid phase

m	Referring to molalities
R	Residual
x	Referring to mol fractions
α	Referring α phase
β	Referring β phase
∞	Referring to infinite dilution
0	Referring to pure component

Dimensionless numbers

$Ar = \dfrac{\Delta\rho \; g \, d_P}{\rho_C (\eta_C / \rho_C)^2}$	Archimedes number
$Bo = \dfrac{w \, H_c}{D_{ax}}$	Bodenstein number
$E\ddot{o} = \dfrac{\Delta\rho \; g \, d_h^2}{\sigma} = \dfrac{\mathrm{We}}{\mathrm{Fr}}$	Eötvös number
$Fo = \dfrac{D \, t_e}{d_p^2}$	Fourier number
$Fr = \dfrac{U_h^2}{g \, d_h}$	Froude number
$K_F = \dfrac{\sigma^3 \; \rho_C}{\eta_C^4 \; g} \dfrac{\rho_C}{\Delta\rho}$	Liquid number
$K_{HR} = \dfrac{1 + \eta_C / \eta_P}{1 + 2/3 \, \eta_C / \eta_P}$	Hadamard–Rybcynski factor
$Nu_c = \dfrac{\eta_c}{\rho_c \; D_{ij}^c}$	Nusselt number
$Pe = \dfrac{w \, l}{D}$	Peclet number
$Re_S = \dfrac{V_S \; d_P}{(\eta_C / \rho)}$	Slip Reynolds number
$Re_\infty = \dfrac{w_T \; d_p}{(\eta_C / \rho_C)}$	Single droplet Reynolds number
$Re_H = \dfrac{U_h \; d_h}{(\eta_d / \rho_d)}$	Hole Reynolds number
$Sc = \nu / D$	Schmidt number
$Sh = \dfrac{k \, d}{D}$	Sherwood number

$We = \dfrac{U_h^2 \, \rho_d \, d_h}{\sigma}$	Weber number
$We_{m,Cau} \dfrac{\rho_c^{0.8} \, d \, D_R^{1.6} (w^{1.8} - w_{crit}^{1.8}) \eta_c^{0.2}}{\sigma}$	Modified Weber number [Cauwenberg 1995]
$We_R \dfrac{\rho_c \, d \, D_R^2 \, N^2}{\sigma}$	Rotor Weber number

1 Reactive Extraction

1.1 Introduction

Liquid–liquid extraction, leaching (solid extraction), absorption (gas extraction) and extractive distillation have in common the use of a solvent which achieves the separation. According to this, solvent recovery in a continuous solvent circuit is necessary. Usually liquid–liquid extraction operates at ambient or slightly higher temperatures and reextraction may be accomplished by a back wash or thermal strip. The transfer of a dissolved compound (solute) from one phase to another may be enhanced when admixing "salting out" agents or more selectively achieved by using "reacting" compounds dissolved in the solvent phase. The use of liquid ion exchangers, diluted in a solvent, has given rise to new approaches in the field of hydrometallurgy, in the chemical and bio-chemical industry and in environmental applications. An excellent survey of progress in these fields is given in the proceedings of the International Solvent Extraction Conferences [ISEC '71 – '99].

The history of extraction dates back in "life-science" applications to about 3500 BC [Blaß et al. (1997)]. An extraction pot operating quite similar to a modern Soxleth apparatus was found 250 km north of Bagdad (see Fig. 1.1). Solid feed materials from vegetables or animals were extracted under reflux with hot water or oil. The production of the first perfumes, pharmaceutical active oils and waxes was documented by a Sumerian text of 2100 BC (see Fig. 1.2). In the medieval age the greatest improvement was with the new solvent ethanol or in the field of hydrometallurgy when using mineral acids or amalgams for metal extraction and raffination.

The major step forward was in the late 19th century with the development in thermodynamics (distribution law by Nernst (1891)) and apparatus design. The first liquid–liquid extraction column (Fig. 1.3) was patented by Pfleiderer (1898) containing all the features (stirrer to enhance mass transfer, an inlet distributor, coalescence aids, countercurrent flow) of modern columns. Jungfleisch and Berthelot (1872) also detected a distribution of metal salts between water and an organic phase which was the first application of metal salt extraction. In industry metal salt extraction started with uranium production on the "Manhattan"-project around 1940 [Coleman & Leuze (1978)]. This new method also found wide commercial applications with some rare metals like vanadium, zirconium, hafnium, niobium and tantalum which were pyrometalurgically diffult to produce at high quality. The break-through was in the sixties when General Mills (Henkel) introduced their LIX[1] chemicals [Power (1971)] for selective copper recovery from acidic leach solutions. Nowadays, the use of liquid ion exchangers has expanded to virtually all metals in the mining and environmental business, but also to the fields of extraction of organic and inorganic acids, organic chemistry intermediates, pharmaceuticals, etc. Such a reactive extraction process [Marr & Bart (1982)] can be used for

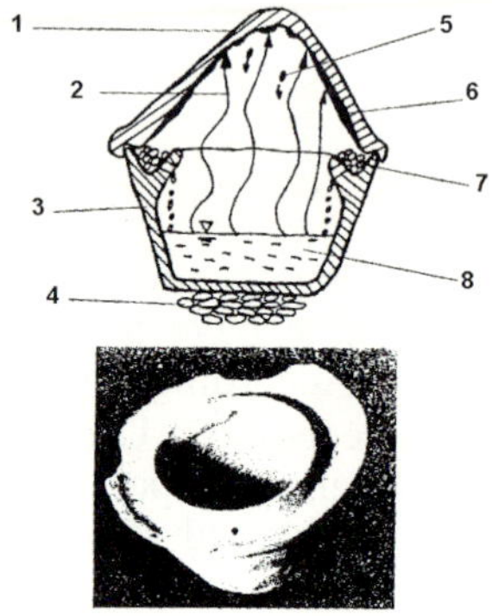

Fig. 1.1: Extraction pot: 1 = cooling cap, 2 = vapour, 3 = extracting pot, 4 = heating, 5 = condensate droplet, 6 = condensate film, 7 = feed material, 8 = solvent (oil or water)

Fig. 1.2: Sumerian text: Purify and cut the skin of a water snake; pour water over the AMASH DUBKASAL, roots of myrtle, alkali powder, barley and powdered pine tree resin; boil mixture; decant the water; wash the remaining liquid, mix it with olive oil; add SAKI [Holmyard (1957)]

[1] *LIX*: liquid ion exchanger

- separation or purification,
- enrichment or
- conversion of a salt.

Enrichment problems are mainly observed in environmental applications when removing toxic materials (e.g. heavy metals) from effluents. The separation and purification applications in industrial practice are numerous for metals [Ritcey & Ashbrook (1979)] or other solutes [Lo et al. (1983)] reviewed in several handbooks [Thornton (1992)]. The conversion method is very specific and offers the possibility to generate a product which can be easily used further on. An example from the mining business is when converting sodium tungstate to ammonium tungstate in a liquid anion extraction process. After the crystalline product is thermally cracked, ammonia is recycled and WO_3 powder produced [MacInnis & Kim (1983)]. Another example is the extraction of citric acid from fermentation broths. When stripping with hot water from the liquid ion exchanger, citric acid is formed instead of calcium citrate from an alternative precipitation process [Weatherly (1992), Bauer et al. (1989)]. However, if volatile products like acetic acid, furvoral, etc. are extracted, a distillative recovery will give the desired product.

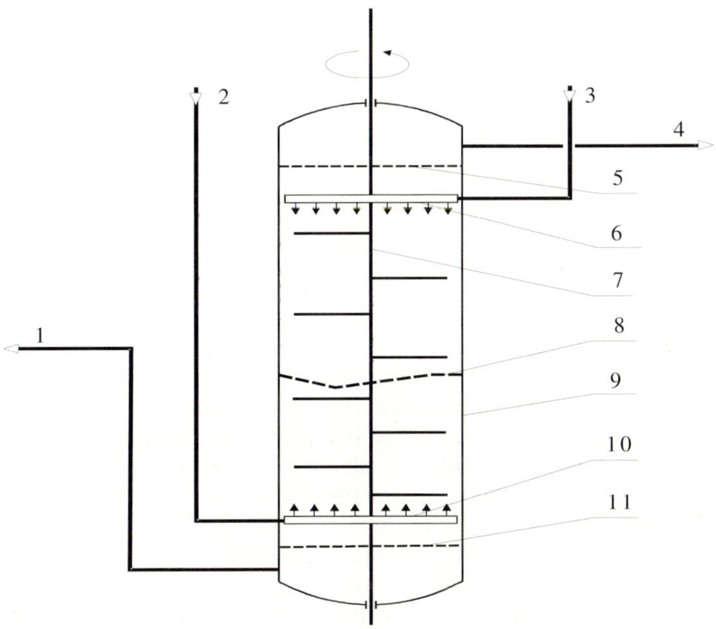

Fig. 1.3: Extraction column: 1 = raffinate, 2 = feed, 3 = solvent, 4 = extract, 5 = coalescence aids, 6 = liquid distributor, 7 = stirrer, 8 = principle liquid–liquid interface, 9 = cylindrical column wall, 10 = liquid distributor, 11 = coalescence aids [Pfleiderer (1898)]

1.2
Physical Extraction Systems

Liquid–liquid extraction operations, phase equilibria, equipment and extraction methods of physical extraction systems are extensively reviewed in the literature [Robbins & Cusack (1997)] and therefore only a brief introduction to the principles is given here. The physical extraction process is based on the partial miscibility of at least three components. Fig. 1.4 depicts a ternary liquid–liquid mixture in which the binodal line separates the phase region from the homogeneous region. The node lines connect the two phases that coexist when phase splitting occurs. Between the binodal line and the spinodal line lies a metastable area while the area surrounded by the spinodal line is completely non-stable. Both lines meet in the critical point. The extract phase and the raffinate phase also meet at this point.

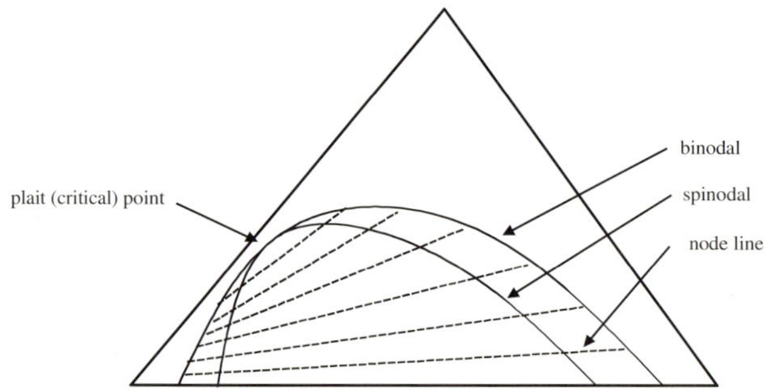

Fig. 1.4: Liquid–liquid ternary phase diagram

In the simplest extraction system we have a mixture of two compounds, *A* and *B*, which are to be separated. This can be done by extracting with a carefully selected solvent, *S*, in which sample *A* (the solute) easily dissolves but *B* (the nonsolute) does not. The solvent phase leaving the column is the extract and the other liquid phase leaving the column is the raffinate. In the countercurrent extraction process there is a light and a heavy phase; one phase is dispersed in the other continuous phase. Which phase is dispersed is an important choice in the design of the process. The location of the principle interface depends on it as does the mass transfer. Finally, the solvent is to be recovered from the extract for recycling. To explain the extraction process, we will focus first on liquid–liquid equilibria.

Analogous to vapour–liquid equilibria we can use *K*–values for liquid–liquid equilibria defined as the ratio of the fugacity coefficients for each component. We denote the extract phase with *E* and raffinate with *R* (which phase will rise depends on the densities). Using the same liquid model for both phases we obtain (in fugacities or activities):

$$K_i = \frac{\phi_i^E}{\phi_i^R} = \frac{a_i^E}{a_i^R} \tag{1.1}$$

Here we have defined the *K*–value in terms of the extract and the raffinate phase, so that the *K*–value is independent of which phase is the light and which phase is the heavy phase. If ideal solutions are considered, the *K*–value equals the distribution coefficient D_i which is defined as the ratio of the concentrations in the liquid phases.

$$D_i = \frac{c_i^E}{c_i^R} \tag{1.2}$$

The distribution coefficient gives a measure for the solubility of component *i* in the extract and the raffinate phase (a larger *D* means a higher solubility in the extract phase). In an analogy to the relative volatility for distillation processes in liquid–liquid extraction, a similar separation factor, the selectivity, can be defined as the ratio of distribution coefficients of the species *i* and *j*:

$$S_{ij} = \frac{D_i}{D_j} \tag{1.3}$$

In a ternary system we can have various types of liquid–liquid equilibria (see Fig. 1.5). In type I we have only one binary subsystem with a miscibility gap. In type II equilibria we have two binary subsystems in which phase splitting occurs. In such a system we can have either two heterogeneous regions or in the case of strong demixing these regions can link to only one two-liquid phase region in the ternary diagram. Of course, the consequences for the extraction operation of such a change can be large. Type III has three binaries that are only partly miscible.

Liquid–liquid equilibria can show a strong temperature dependence. Fig. 1.6 shows various cases. If there is a point at which the two-liquid phase region disappears we have a critical solution temperature point (CST). Dependent on the system we can then have either an upper CST (UCST) or a lower CST (LCST), or both as indicated in the fourth picture of Fig. 1.6. The thermodynamics that describe such complex phase behaviour will be discussed in Chapter 2.

A typical extraction process is depicted in Fig. 1.7. On top of the extraction column the solvent is added and flows countercurrent to the feed and an extract reflux which enters the extractor at the bottom. On top of the column the raffinate leaves the system while the extract that is obtained at the bottom is sent to a stripper, usually a distillation column, after being preheated. The top product of this column is the desired extract product from which a certain amount is fed back to the extractor as extract reflux. The recovered solvent leaves the stripper at the bottom and is cooled down to operating temperature before entering the extractor again.

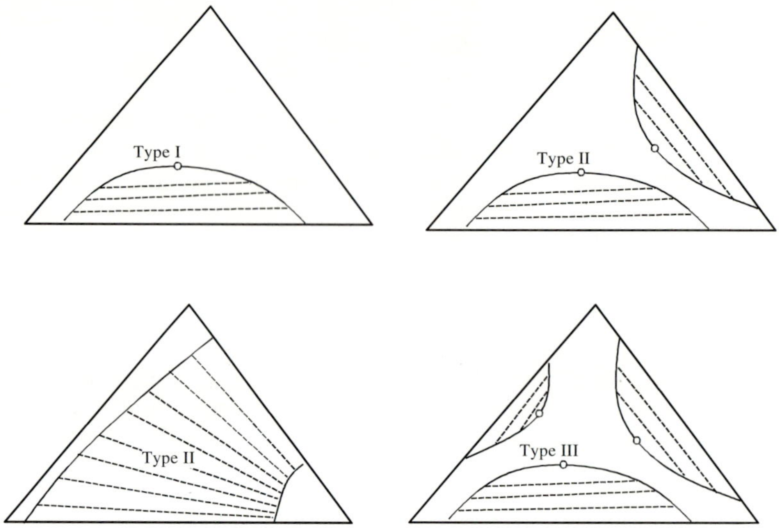

Fig. 1.5: Liquid–liquid equilibria

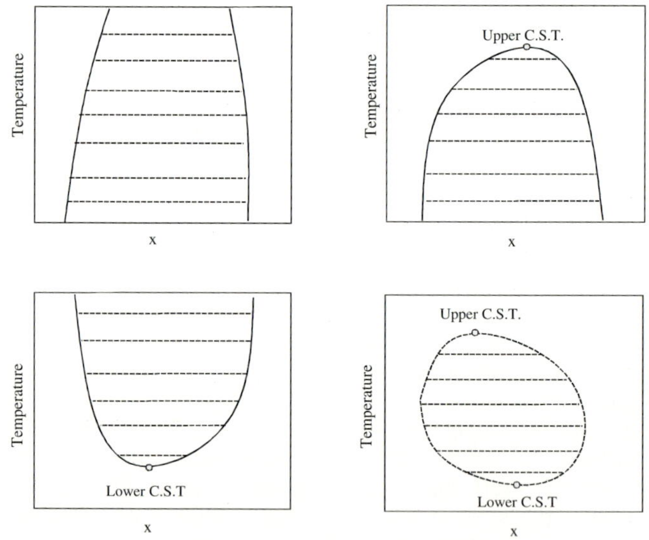

Fig. 1.6: Temperature dependence of liquid–liquid equilibria

The effect of an additional extract or raffinate reflux is shown in Fig. 1.8. If pure solvent S and feed F enter a single-stage liquid–liquid extractor and reach equilibrium, the leaving raffinate and extract compositions will be R and E, respectively. If a reflux of pure extract is fed to the extractor as well, the total feed composition changes and the purity of the leaving extract will increase.

This further purification can be achieved only at the expense of a cycle stream and additional costs associated with it.

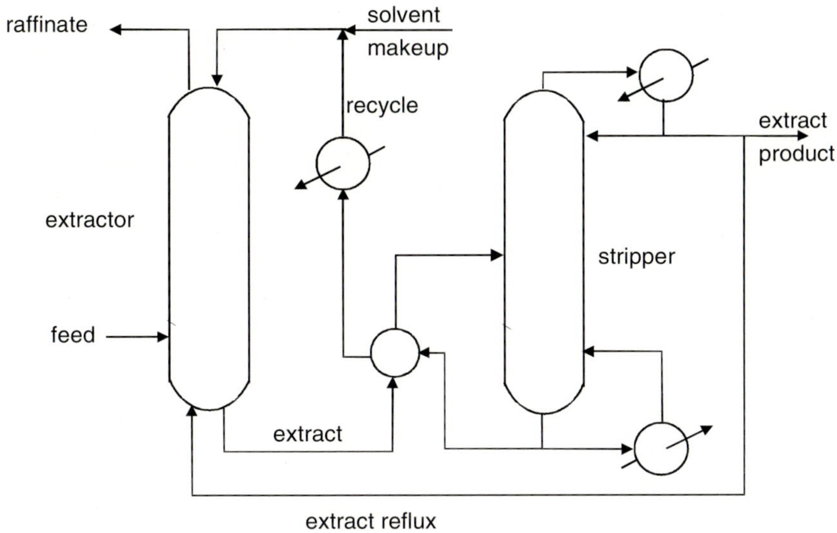

Fig. 1.7: Extraction process

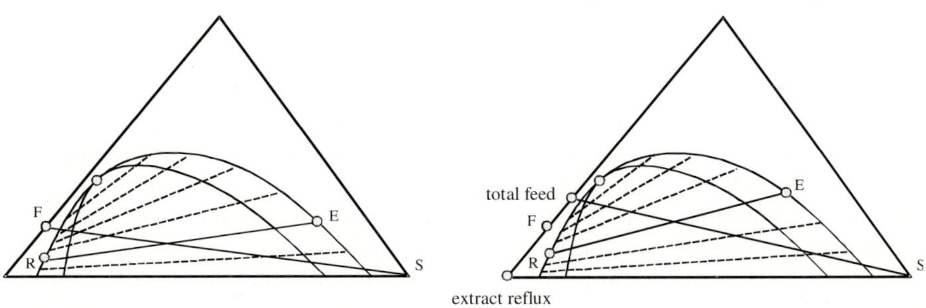

Fig. 1.8: Equilibrium extraction stage

If the leaving fluxes of a single-stage extractor are not in equilibrium the separation becomes worse. Then the compositions of the coexisting phase do not lie on the binodal line as indicated in Fig. 1.9. This deviation can be described by the Murphree stage efficiency which relates the theoretical and the practical stage. In liquid–liquid extractors the Murphree stage efficiencies can vary in the range of 5% to 50% and prediction of these is often difficult. With the introduction of such efficiencies several problems arise. Due to the different mass transfer behaviours of the molecule species, a stage efficiency has to be defined for each component separately. In multicomponent mixtures they can vary from $-\infty$ to ∞ caused by multicomponent interactions between the molecules. Therefore the nonequilibrium model should lead to physical more meaningful solutions since equilibrium is only assumed at the phase interface. However, the application of the nonequilibrium model in commercial flowsheeting programs (ASPEN Company, ChemSep Company) for extraction problems is limited. It is clear that cases where we find diffusion of components against their gradient exist in liquid–liquid extraction [Taylor & Krishna (1993), Nitsch & van Schorr (1983)], which can only be treated with a nonequilibrium model. A discussion of this is presented in Chapter 3.

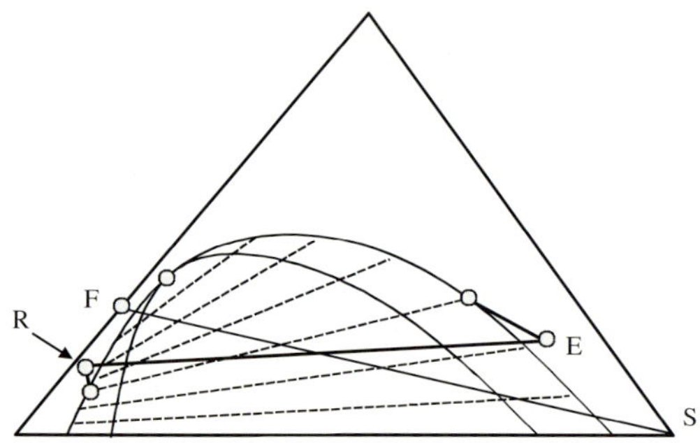

Fig. 1.9: Nonequilibrium extraction stage

Actual extraction processes have extract and/or raffinate treatments to recover entrained (slipped) phases. For example there might be a reasonable amount of solvent present in the raffinate which needs to be separated and recycled. If only the solvent dissolves in water, a simple water-wash might work as recovery separation. Quite often used is air flotation or coalescing aids [Henschke (1995), Miller et al. (1996)]. Coalescers are mats, beds, staggered metal sheets, layers of porous or fibrous solids with surfaces especially suited to reduce the slip of phases. All these extra auxiliary separation techniques lead to an increase of the investment costs and the operating costs of the liquid–liquid extraction unit.

The selection of an appropriate solvent is key to successful separation by liquid–liquid extraction. Below is a list of the various solvent-selection criteria. Some of these are essential for the separation while others are desirable properties which will improve the separation and/or make it more economical. The solvent selectivity, recoverability and a density difference with the raffinate which is large enough are essential. Some of the requirements on the solvent will contradict and a compromise may be necessary.

Selectivity – a high value of the separation factor S_{ij}, enables fewer stages to be used. Extract or raffinate purities can be improved by using more solvent (changing the solvent to feed ratio). If the feed is a complex mixture where multiple components need to be extracted, group selectivities become important.

Capacity – a high value of the equilibrium ratio, K, indicates a high solvent capacity for solute and permits to lower solvent to feed ratios. Since the K–value is a function of temperature, a temperature gradient over the extractor might be used to increase the capacity (higher temperatures tend to increase the capacity but lower selectivity). Also, addition of other components (anti-solvents) might influence the K–value. Often a compromise between selectivity and capacity must be found. The use of an extract or raffinate reflux can also improve the purity of either product.

Recoverability of the solvent – recovery of the solvent should be easy, preferably by means of a simple flash or a flash followed by a stripping column. This is accomplished by using a solvent with much higher or lower boiling temperature than the component to be extracted. In some cases more effort might be warranted and the extract is followed by an extractive stripper and a distillation.

Density – the density difference must be large enough to ease the settling of the liquid phases. Higher density differences permit higher equipment capacities. Systems with very low density differences require expensive centrifugal extractors.

Viscosity and melting point – high viscosities reduce the mass transfer efficiency and lead to difficulties in pumping and dispersion. Low viscosities also benefit rapid settling and capacity (the more viscous phase is usually dispersed). Column temperatures are usually determined by the viscosities. The melting temperature of the solvent should preferably be lower than ambient for ease of handling.

Insolubility of solvent – the mutual solubilities of nonsolute and solvent should be low. If this is not the case, an additional separation step is necessary to recover the solvent from the raffinate.

Interfacial tension – high interfacial tension permits a rapid settling due to an easier coalescence, permitting higher capacities. Low interfacial tension facilitates the phase dispersion, large interfacial areas can be easily achieved and hence higher separation efficiencies, but may require large volumes for phase separation. An interfacial tension that is too low leads to emulsification.

Toxicity and flammability – for food processing only nontoxic solvents will be taken into consideration. In general, any hazard associated with the solvent will require extra safety measures.

Corrosivity – corrosive solvents increase equipment cost but might also require expensive pre- and post-treatment of streams. Such a problem can be alleviated somewhat by dispersing the corrosive solvent.

Thermal and chemical stability – it is important that the solvent should be thermally and chemically stable as it is recycled. Especially it should resist breakdown during the solvent recovery in for example stripping columns. Sometimes special measures are required to prevent solvent degradation (for example for furfural, a temperature limitation, nitrogen blanketing and feed de–aeration are required).

Availability and cost – solvent should be readily available. It is not the price of the solvent that is important, but the annual cost due to the inevitable operation losses.

Environmental impact – the solvent should not only be compatible with upstream and downstream process steps, but also with the environment (minimal losses due to evaporation, solubility and entrainment).

1.3 Chemical Extraction Systems

In order to enhance the selectivity or to extract ionic solutes, the use of liquid ion exchangers is recommended. They can be applied to a manifold of extraction processes in the chemical industry (e.g. extraction of furfural, organic and inorganic acids, etc.), biochemical and pharmaceutical products (e.g. penicilline, amino acids, etc.), in hydrometallurgy (e.g. mining of metals) and in all related environmental applications. The latter area is especially attractive, since liquid ion exchangers react with good selectivity and high performance at low solute feed concentrations.

For practical purposes, the ion exchanger is diluted preferably in a non-aromatic, high boiling (b.p. ≈ 500 K) diluent, which is immiscible with water. This gives the organic phase the needed physical properties (high interfacial tension, low viscosity, low density), since most liquid–ion exchanges are highly viscous or even solid. At very high solute loadings a split of the organic phase in a solvent-rich and a solvent-less fraction may occur, especially when using aliphatic diluents. Such a liquid three-phase system is not envisaged in a solvent extraction process. In these cases a modifier, usually a long chain alcohol (e.g. isododecane) is added to increase the solubility properties of the solute-ion exchanger complex. Aromatic diluents have more solvating power and will avoid three-phase formation but are not always the first choice in industrial application, as is in the food industry due to toxicity or in the nuclear industry due to, for example, nitrification in strong acidic media. However, the admixture of a modifier to the organic phase is not always favoured since the solubility is usually different from the other organic components. With time the composition of the organic phase will gradually alter and change its designed properties.

The role of the diluent in solvent extraction is always a hot topic. Coleman and Leuze (1978) reported a two-fold increase of the distribution coefficient with a two-fold dilution of the ion exchanger. The influence is therefore not only on physical properties and hydrodynamics but also on mass transfer and extraction equilibria [Diamond (1967), Ritcey & Lucas (1974), Högtfeld (1976), Blumberg & Gai (1977), Ashbrook et al. (1979), Khalifa et al. (1992), Tian et al. (1992), Shelley & Quan (1992), etc.]. A detailed review on liquid ion exchangers, diluents and modifiers which form a reactive extraction system is given in several textbooks [Lo et al. (1983), Thornton (1992)]. In general the variety of a diluent is rather limited (aliphatic, naphtenic or aromatic ones) since in contrast to the physical extraction systems the choice of the ion exchanger is the decisive step in process design.

A short review on the principles involved is given by Marr and Bart (1982) with the different liquid ion exchangers commercially available which rely on three basic extraction mechanisms:

- anion exchange,

- cation exchange and
- solvation

as is discussed below.

An anion exchange of an anionic zinc complex (e.g. from an acidic chloride brine) is as follows (a bar indicates the organic phase species):

$$2\overline{R_4NHCl} + ZnCl_4^{2-} \longleftrightarrow \overline{(R_4NH)_2ZnCl_4} + 2Cl^- \tag{1.4}$$

The quarternary R_4-alkyl-substituted ammonium chlorides are commercially available anion exchangers and can be stripped with an excess of chloride, hydroxide etc. and thus the solute is purified in the reextraction step. Primary, secondary (both water soluble, less used) and tertiary amines are only stable in acidic aqueous media, since hydroxide ions destroy the complex:

$$\overline{R_3NHCl} + OH^- \longleftrightarrow \overline{R_3N} + Cl^- + H_2O \tag{1.5}$$

However, volatile anions like acetate, formiate, etc. can also be removed and stripped by distillation which yields the free tertiary amine, R_3N, similar to equ. (1.5). The changes in counter ion concentration and temperature give rise to a reversible extraction process according to equ. (1.4). Generally, the selectivity of anion exchangers is not always satisfactory, which gives rise to new developments on host–guest–ligands, which take the size and shape of the solute into account [Gloe et al. (1999)]. However, these complexes are then of high molecular weight and therefore less soluble in diluents like kerosene. A straightforward application in liquid extraction is not easily feasible but an adsorption and immobilisation with solvent-impregnated resins (see Chapter 4) may be the proper choice for technical applications.

The cation exchange mechanism in a slightly acidic medium is as follows:

$$2\,\overline{D2EHPA} + Zn^{2+} \longleftrightarrow \overline{Zn(D2EHPA)_2} + 2H^+ \tag{1.6}$$

Here zinc is extracted with *D2EHPA*[2] in its *H*-form and two protons are set free. This causes a pH shift during extraction, which can be avoided if the ion exchanger is, for example, in the sodium form. For the reextraction usually strong mineral acids (preferable H_2SO_4) are used. Besides aryl and alkyl substituted phosphoric acids are phosphonic and phosphinic acids and their thio-forms available. The latter ones are strong extractants and applicable at a pH value of the feed lower than 1 but difficult to strip. In general, the feed pH with these reagents is between one and five, since at a higher pH value most metals precipitate. Carboxylic based ion exchangers are seldom used due to their high water solubility. As a special feature chelating ion exchangers are available (e.g. LIX (Henkel Company), Acorga (ICI Company) reagents), which are preferably used for copper recovery from low grade ores. A size specific host–guest complexation leads to a selective copper recovery from iron containing brines. As can be seen in Fig. 1.10, the nitrogen additionally chelates the copper ion and the developing new six-ring structure is only stable with copper as solute.

---

[2] *D2EHPA*: di(2-ethylhexyl) phosphoric acid

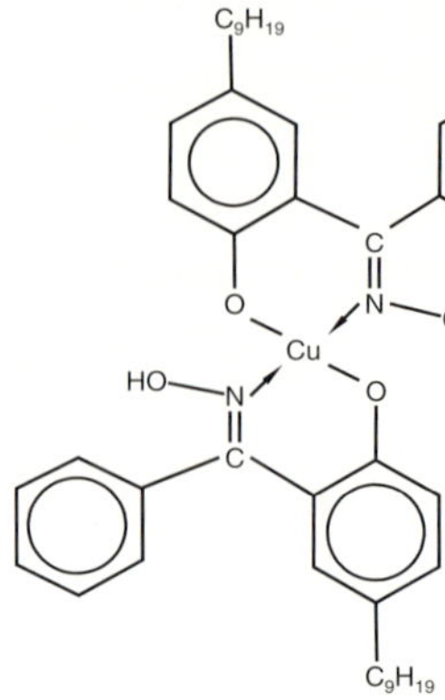

Fig. 1.10: Copper chelate structure

An extremely non-stoichiometric behaviour occurs with neutral agents or solvating ion exchangers:

$$ZnCl_2 + x\overline{TBP} \longleftrightarrow \overline{ZnCl_2(TBP)_x} \tag{1.7}$$

TBP[3] has a very high solvating power for neutral substances, e.g. acids, salts, and will extract these. Higher temperatures and pure water will achieve re-extraction. Most of the solvating exchangers (alkyl substituted phosphates, phosphonates, phosphine oxids, ethers, etc.) can be used undiluted due to appropriate physical properties. Carbon based solvating ion exchangers (e.g. *MIBK*[4]) are usually more water soluble but are volatile and can be separated from the product by distillation. As can be seen, a recovery by distillation of the solvent is the general case with physical extraction systems, so the physical extraction chemicals are a subgroup within the neutral ion exchangers. However, nowadays ionic liquids (e.g. butylmethylimidazolium hexafluorophosphate) are novel media for reactive extraction [Huddleston et al. (1998)] with the prevalence of high coulombic forces resulting in the relative absence of any significant vapour pressure.

A special system is an equimolar mixture of cation and anion exchanger which gives a "mixed" extraction system that allows the extraction of salts or acids in their dissociated forms:

$$2\overline{R_3N} + 2\overline{D2EHPA} + Zn^{2+} + 2Cl^- \leftrightarrow \overline{(R_3NH)_2Cl_2 \cdot Zn(D2EHPA)_2} \tag{1.8}$$

Reextraction can then be accomplished either by shift of temperature, aqueous ionic strength or acidity/basicity. The protons set free by the cation exchanger (see equ. 1.6) are consumed by the amine. All kinds of liquid ion exchangers can be mixed together in order to generate synergistic effects. This means that the effect of a mixture gives a non-linear improvement in regard to the single

[3] *TBP*: tri-butyl-phosphate

[4] *MIBK*: methyl isobutyl ketone

systems. Basically the effect results from an improvement to solvate the new ion exchanger solute complex. Such synergistic behaviour is also reported with extraction kinetics when Henkel improved their first commercial copper extraction reagent *LIX64*[5] with an admixture of a small amount of *LIX63*. It markedly enhanced the kinetics of the new reagent *LIX64N* [Kordosky (1973)] which was a milestone and the commercial start to extract base metals on a big scale. In general one can find numerous references to synergistic effects reviewed in all the solvent extraction textbooks.

1.4 Process Development and Design

The process development of a reactive extraction process is primarily bound to the availability of a certain type of ion exchanger. Usually in a screening procedure the separation factors will be determined experimentally [Ritcey & Ashbrook (1979) and a process flowsheet will be derived. The practical handling and design of a reactive solvent extraction process is given in several reviews and books [Cox (1992), Cox & Flett (1983), Godfrey & Slater (1994), Lo et al. (1983), Ritcey & Ashbrook (1979), Rydberg et al. (1992), Thornton (1992)]. A simple example is with cation exchange. The law of mass action for equ. (1.6) is

$$K_{Zn} = \frac{\overline{[Zn\,(D2EHPA)_2]}\;[H^+]^2}{\overline{[D2EHPA]}^2\;[Zn^{2+}]} \tag{1.9}$$

Converted into the logarithmic form with the definition of the distribution coefficient for zinc (equ. (1.2)) the equation can be rewritten as:

$$\log D = \text{const.} + 2\ \text{pH} \tag{1.10}$$

For a divalent metal the change in one pH-unit will change the distribution coefficient 100 fold. The $pH_{0.5}$ value is that where D equals one and characterises the separation sequence of different ions. For an extract purity of 99%, log $D = 2$, and for a 1% loss of raffinate, log $D = -2$. This is depicted in Fig. 1.11 where $4/n$ is the distance of the distribution isotherms of the two metals to achieve that separation. The lower the oxidation state, n, of the metal ion is, the lower is the inclination of the isotherms and it is the more difficult to achieve a sharp separation. If the required separation cannot be achieved during extracting, the pH value and aqueous concentrations can be adjusted accordingly in an additional scrubbing step to meet this criterion [Ritcey, Ashbrook (1979)] (compare Fig. 1.12).

[5] *LIX64:* liquid ion exchanger 1964

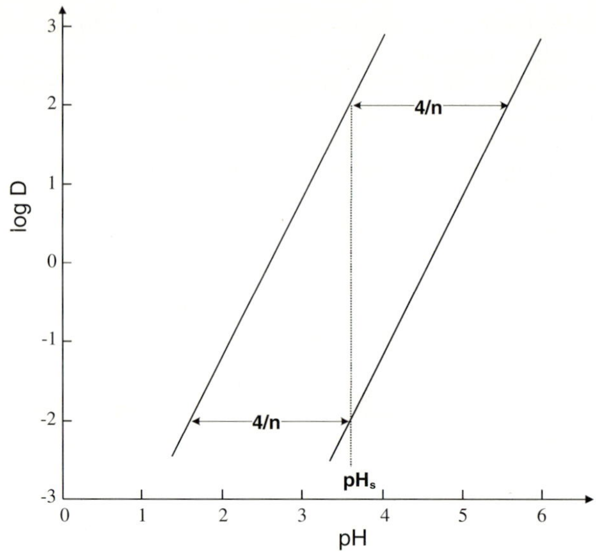

Fig. 1.11: Separation pH for two metal ions

However, there are some striking differences in respect of a physical extraction circuit. A temperature swing for stripping (see Fig. 1.7) is usually not feasible since neither the extracted products nor the organic solvent phase is volatile. The reextraction is performed with a "chemical" swing using different ionic strength, concentrated acids or salts as stripping agents. As likewise in physical extraction, an extract reflux is fed back to a scrubbing section (see Fig. 1.12) to improve the product quality. The impurities then leave with the raffinate stream (5). The scrub conditions are sensitive to temperature, concentrations and pH, which are different from the extraction and reextraction stage. Very often pH (stream 10) is adjusted to meet special criteria. Sometimes the liquid ion exchanger is regenerated (4) and converted to a certain salt form. For example, a cation exchanger in its acidic H-form may be converted to its neutral sodium form by neutralisation (stream 12) prior to its use in the extraction step. However, if extraction is in sulphate and reextraction is in chloride media, the scrub and regeneration step may be a simple water wash in order to reduce entrainment and hinder a mutual contamination of the different aqueous media. As can be seen, a reactive extraction circuit is very flexible and offers a manifold of options for process development.

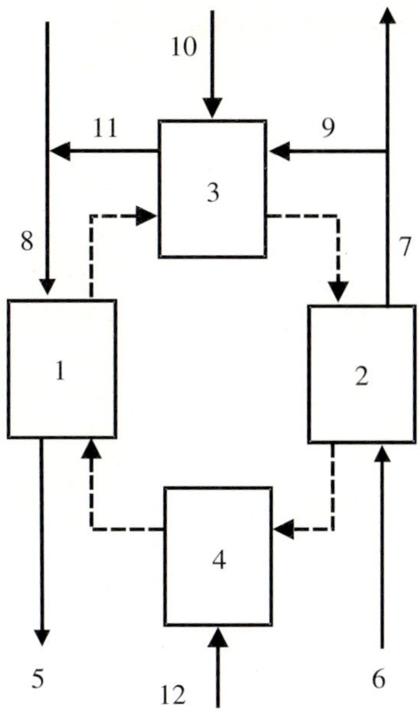

Fig. 1.12: Reactive extraction circuit: 1 = extraction, 2 = reextraction, 3 = scrubbing, 4 = regeneration, 5 = raffinate, 6 = stripping solution, 7 = product solution, 8 = feed solution, 9 = product reflux, 10 = pH adjustment, 11 = aqueous scrub raffinate, 12 = neutralisation, (dotted line: organic phase).

From an engineering point of view the process design relies on four major steps (see Fig. 1.13) with the knowledge of:

- reactive phase equilibria,
- microkinetic parameters,
- macrokinetic influences and
- apparatus selection and design.

The reactive equilibria are either determined in a stirred vessel or in a shaking funnel. The thermodynamic models for reactive liquid–liquid equilibria are then discussed in Chapter 2. The microkinetic parameters can be obtained only experimentally in a Lewis-type stirred cell where a reaction controlled regime prevails and true kinetic parameters (without any diffusional contribution) can be derived. The effective mass transfer in a dispersed phase droplet is then a superposition of kinetic, diffusional and turbulent contributions. As will be seen in Chapter 3, the correlations for these mixed regimes show big deviations and an experimental verification is highly recommended. The experimental determination is either in a rising droplet apparatus (for short residence times) or in a Venturi-tube (for long residence times). The influence of inert salts and surfactant layers on mass transfer and their experimental characterisation is discussed, too, in examples of zinc (cation exchange) and citric acid (anion exchange). The final Chapter 4 is then related to apparatus' design and current developments in reactive liquid extraction.

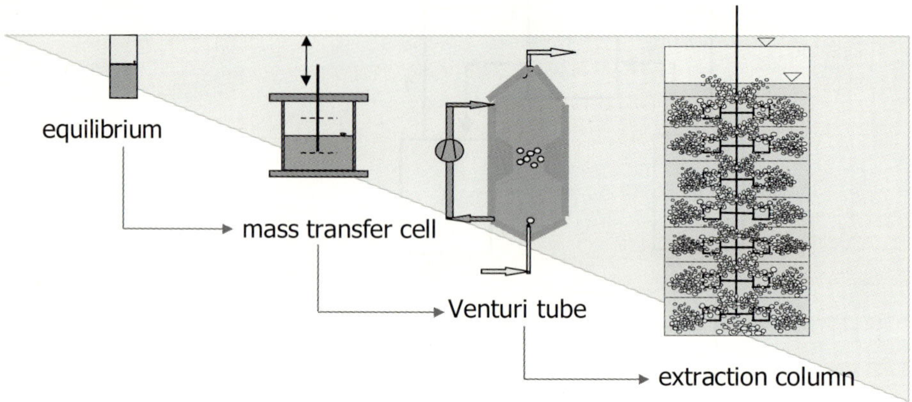

Fig.1.13: Process development scales

2 Liquid–Liquid Phase Equilibria

2.1 Equilibrium Conditions

The design of an extraction process and the layout of an extraction apparatus is based on thermodynamics, hydrodynamics, and mass transfer. Among these, the thermodynamic data for the distribution of the solute between the two phases is the most important. There are several available source books on liquid-liquid equilibria (LLE), e.g. for organic solutes [Wisniak & Tamir (1980), Sorensen & Arlt (1979–1980)]. The most commonly used correlations for LLE are also implemented in simulation tools like ChemSep®, ASPEN®, HYSIM®, etc. Unfortunately, the available data may be incomplete or unreliable which then makes laborious experimental work necessary.

In heterogeneous systems the phase rule specifies the degree of freedom in which the number of intensive variables can be altered independently without changing the system. The degrees of freedom, F, are related to the number of phases, N_p, and components, N_i, after Gibbs as to:

$$F = N_i - N_p + 2 \tag{2.1}$$

A two component, two phase system has two degrees of freedom, that is two intensive variables such as temperature and pressure specify the composition completely. The phase rule says that at a given temperature and pressure when only a single phase is present, there will be one degree of freedom and we can choose the composition of this phase, made up from two components, at will. This does not preclude the phenomenon of saturation, for which beyond a certain amount of solute in a liquid mixture, a new phase appears which may be either solid or liquid. However, at a certain temperature and pressure there are then no longer any degrees of freedom.

In a heterogeneous system with two phases, for example, the conditions for the thermodynamic equilibrium may be written as follows:

- thermal equilibrium

$$T^G = T^L \tag{2.2}$$

- mechanical equilibrium

$$p^G = p^L \tag{2.3}$$

- physical equilibrium

$$\mu_i^G = \mu_i^L \tag{2.4}$$

- reaction equilibrium

$$\sum_{i=1}^{n} \nu_{i,R}\, \mu_i^L = 0 \qquad R = 1,2,\ldots,k \tag{2.5}$$

The chemical potential (also known as partial molar Gibbs free energy, g_i,) of compound i is μ_i which is related to the Gibbs free energy G^α of any phase, α, by:

$$g_i^\alpha \equiv \mu_i^\alpha = \left(\frac{\partial G^\alpha}{\partial n_i^\alpha}\right)_{T,p,n_j^\alpha} \tag{2.6}$$

An alternative and sometimes more useful criterion for the chemical equilibrium is that at constant pressure and temperature the Gibbs free energy function of the entire system of ω-phases shows a minimum:

$$\mathrm{d}\left(\sum_{\alpha=1}^{\omega} G^\alpha\right)_{T,p} = 0 \tag{2.7}$$

In addition to the physical phase equilibrium in systems with chemical reactions, the reaction equilibrium must hold as given in equ. (2.5). In a n-component system with k reactions $\nu_{i,R}$ is the stoichiometric coefficient of compound i in the reaction R. An example of such a reaction equilibrium is with the dissociation of $NaCl$:

$$NaCl \leftrightarrow Na^+ + Cl^- \tag{2.8}$$

with k = 1, n = 3 and ν_{NaCl} = -1. In order to evaluate the physical and reaction equilibrium according equs. (2.4) and (2.5) the chemical potential of species i is required. Following Lewis, the chemical potential in a mixture is the sum of a reference state, μ_i^o, and a concentration dependent correction term.

$$\mu_i = \mu_i^0 + RT \ln a_i \tag{2.9}$$

The reference state for gases is the pure ideal gas at temperature T and standard pressure p^* = 0.1 MPa. The activity, a_i, is the ratio of the fugacity of component i in the mixture to its fugacity at the standard state. The deviation of the partial pressure from the fugacity is expressed with the fugacity coefficient, φ_i^G, which is unity for an ideal gas.

$$a_i^G = \frac{f_i^G}{f_i^0} = \frac{p\, y_i\, \varphi_i^G}{p^*} \tag{2.10}$$

The chemical potential in the gas phase is then:

$$\mu_i^G = \mu_i^0 + RT \ln\left(\frac{p\, y_i\, \varphi_i^G}{p^*}\right) \tag{2.11}$$

In analogy to equ. (2.9) the chemical potential in the liquid phase is then:

$$\mu_i^L = \mu_i^0 + RT \ln\left(\frac{p\, x_i\, \varphi_i^L}{p^*}\right) \tag{2.12}$$

The fugacity in both phases can be obtained from an equation of state (EOS) when the system is at high pressures or when any of the components are close to or above the critical point. The most commonly used cubic EOS are the Peng–Robinson (1976) and the Soave–Redlich–Kwong [Soave (1972)] equation. At moderate or low pressures, where the liquid phase is incompressible, the use of Gibbs excess energy models (g^E models) is handier.

In non-ideal mixtures the Gibbs energy of a mixture can be split into three parts. The first one accounts for the contribution of the pure components, the second one for the mixing effects in an ideal mixture, and the last part for the deviation from ideal behaviour of the mixture:

$$G = G^0 + G^{ID} + G^E \tag{2.13}$$

For a binary mixture these single contributions can be written as:

- Contribution of the pure components

$$G^0 = n_1 g_1^0 + n_2 g_2^0 \tag{2.14}$$

- Contribution due to ideal mixing

$$G^{ID} = n_1 RT \ln x_1 + n_2 RT \ln x_2 \tag{2.15}$$

- Contribution due to non-idealities

$$G^E = n_1 RT \ln \gamma_1 + n_2 RT \ln \gamma_2 \tag{2.16}$$

In terms per moles of component *1* and *2* it is

$$\mu_1 \equiv g_1 = g_1^0 + RT \ln x_1 + RT \ln \gamma_1 \tag{2.17}$$

$$\mu_2 \equiv g_2 = g_2^0 + RT \ln x_2 + RT \ln \gamma_2 \tag{2.18}$$

The definition of the activity, a_i, of a solute, i, is completed using a pure liquid component as reference at the pressure and temperature of the solution (symmetric notation):

$$\lim_{x_i \to 1} (a_i / x_i) = \lim_{x_i \to 1} \gamma_i = 1 \tag{2.19}$$

thus

$$\mu_i^L = \mu_i^0 + RT \ln a_i = \mu_i^0 + RT \ln x_i\, \gamma_i \tag{2.20}$$

However, gases and ions can be found only in diluted form in liquids (at least at conditions that we are considering). It is then handier to choose a different reference point, e.g. the condition of the gas or ion at infinite dilution in a solvent (non-symmetric notation):

$$\lim_{x_i \to 0} (a_i / x_i) = \lim_{x_i \to 0} \gamma_i^H = 1 \tag{2.21}$$

thus

$$\mu_i^L = \mu_i^\infty + RT \ln x_i \, \gamma_i^\infty \tag{2.22}$$

For a component that can be described in both ways, there is a relation between these two activity coefficients. The chemical potential must be the same in both descriptions, thus:

$$\gamma_i - \gamma_i^\infty = \exp\left[\frac{\mu_i^\infty - \mu_i^0}{RT}\right] \tag{2.23}$$

The fugacity of a component *i* in a liquid mixture is given by:

$$f_i^L = \gamma_i \, x_i \, f_i^0 \tag{2.24}$$

Neglecting pressure effects for components at subcritical conditions, the reference fugacity, f_i^0, is usually taken as the Henry constant:

$$f_i^0 = H_i = \lim_{x_i \to 0} \left(f_i^L / x_i \right) \tag{2.25}$$

The reference fugacity can be characterised also by the vapour pressure of the pure liquid component:

$$f_i^0 = p_i^0 = \lim_{x_i \to 1} \left(f_i^L / x_i \right) \tag{2.26}$$

At physical equilibrium the chemical potential of every species in each phase must be equal or as Prausnitz (1969) showed can be replaced by:

$$f_i^{(\alpha)} = f_i^{(\beta)} = \ldots \tag{2.27}$$

For gas–liquid equilibria this leads to:

$$\varphi_i^G \, p \, y_i = p_i^0 \, \gamma_i \, x_i \tag{2.28}$$

or

$$\varphi_i^G \, p \, y_i = H_i \, \gamma_i^H \, x_i \tag{2.29}$$

For most systems of interest the fugacity coefficient, φ_i^G, is close to unity. In ideal solutions the activity coefficient is unity and the resulting equations from equs. (2.28) and (2.29) are known as Raoult's law and the Henry's law, respectively.

For liquid-liquid equilibria the fugacities can be calculated with fugacity coefficients or with activity coefficients. Since both phases (α and β) at equilibrium are at the same temperature, the reference states are the same and the equilibrium condition reduces to:

$$\gamma_i^\alpha \, x_i^\alpha = \gamma_i^\beta \, x_i^\beta \tag{2.30}$$

The distribution ratio, K_i, is then

$$K_i^{\alpha/\beta} = \frac{x_i^\alpha\ \gamma_i^\alpha}{x_i^\beta\ \gamma_i^\beta} \tag{2.31}$$

The distribution ratio remains constant if the ratio of activity coefficients is independent of the species concentration, *i*. This is valid in dilute systems and known as Nernst's distribution law, D_i (usually in molar units):

$$D_i = \frac{c_i^\alpha}{c_i^\beta} \tag{2.32}$$

Although the methods of thermodynamics do not allow the prediction of the composition dependence of the chemical potential, the Gibbs–Duhem equation relates the intensive variables T, p and μ:

$$SdT - VdP = \sum_{i=1}^{n} x_i \mathrm{d}\mu_i = 0 \tag{2.33}$$

and with p and T constant then

$$\sum_{i=1}^{n} x_i\ \mathrm{d}\mu_i = 0 \tag{2.34}$$

and with equ. (2.20):

$$\sum_{i=1}^{n} x_i\ \mathrm{d}(\ln \gamma_i) = 0 \tag{2.35}$$

Thus any empirical or theoretical relations representing the composition dependence of the activity coefficient must be a solution of this equation.

2.2 Non–ionic Systems

2.2.1 The Regular Solution Theory

The regular solution concept allows the prediction of phase equilibria from pure component data. It was first developed by van Laar (1910) who derived activity coefficients from critical data using the van der Waals' equation of state. However, the results show strong dependence on the mixing rules applied. An improvement was when Hildebrand and Scott (1950) replaced the two van der Waals' parameters with two new ones, the molar volume V_i and the solubility parameter δ_i of a solute *i*. The regular solution is one for which the excess entropy of mixing is zero. This is in contrast to an athermal solution for which the excess enthalpy is zero.

$$g^{ID} = h^{ID} - T\ s^{ID} \tag{2.36}$$

and

$$g = h - T\, s^{ID} \tag{2.37}$$

thus

$$g^{E} = g - g^{ID} = h^{E} \tag{2.38}$$

The Gibbs excess enthalpy is therefore independent of temperature and thus:

$$g^{E} = h^{E} = RT\,\ln(\gamma_i) = \text{const.} \tag{2.39}$$

Equating the g^E expressions of the regular solution concept and equ. (2.39) gives:

$$RT \cdot \ln \gamma_i = V_i \cdot \varphi_i^2 \cdot (C_{ii} + C_{jj} - 2 \cdot C_{ij}) \tag{2.40}$$

where φ is the volume fraction of the component i in the mixture. It can be calculated from the molar volumes, V_i, of all solutes and the solvent:

$$\varphi_i = \frac{x_i \cdot V_i}{\sum_{j=1}^{n} x_j \cdot V_j} \tag{2.41}$$

C_{ii} denotes the density of the energy of cohesion as:

$$C_{ii} = \frac{\Delta U_i}{V_i} \tag{2.42}$$

ΔU_i is the internal energy of vaporisation and at normal pressures related to the heat of vaporisation, ΔH_{evap}, as follows:

$$\Delta U_i \approx \Delta H_{evap} - RT \tag{2.43}$$

C_{ij} denotes the intramolecular forces between species i and j and is:

$$C_{ij} = \sqrt{C_{ii} \cdot C_{jj}} \cdot (1 - l_{ij}) \tag{2.44}$$

with the binary parameter l_{ij} (<< 1). When introducing the solubility parameter δ:

$$\delta_i = \sqrt{\frac{\Delta U_i}{V_i}} \qquad \left[\frac{\text{cal}^{\frac{1}{2}}}{\text{cm}^{\frac{3}{2}}}\right] \tag{2.45}$$

the activity coefficient reads as:

$$RT\, ln \gamma_i = V_i\, \varphi_j^2\, (\delta_i - \delta_j)^2 - 2\, l_{ij}\, \delta_i\, \delta_j \tag{2.46}$$

Hildebrand and Scott postulate that the binary parameter l_{ij} is approximately zero. Thus activity coefficients can be calculated from pure solvent data and can be expressed at any temperature as far as molar volumes and solubility parameters are known. For multicomponent mixtures the Hildebrand and Scott model is:

$$RT \ln \gamma_i = V_i \left(\delta_i - \overline{\delta}\right)^2 \quad \text{with } \overline{\delta} = \sum_{k=1}^{n} \varphi_k \, \delta_k \tag{2.47}$$

A Flory–Huggins correction considers effects of different molar volumes:

$$\ln \gamma_i = \frac{V_i}{RT} \left(\delta_i - \overline{\delta}\right)^2 + \ln\left(\frac{V_i}{V_m}\right) + 1 - \frac{V_i}{V_m} \quad \text{with } V_m = \sum_{k=1}^{n} x_k \, V_k \tag{2.48}$$

These effects exceed 10% if the molar volumes of the solutes differ more than 50% with respect to the solvent, V_m. The use of this correction is recommended.

As a reference state is the activity coefficient at infinite dilution where the mole fraction of the solvent x_s approaches unity:

$$\lim_{x_S \to 1} \overline{\delta} = \delta_s \quad \text{and} \quad \lim_{x_S \to 1} V_m = V_s \tag{2.49}$$

and thus:

$$\ln \gamma_i^{\infty} = \frac{V_i}{RT} \left(\delta_i - \delta_s\right)^2 + \ln\left(\frac{V_i}{V_s}\right) + 1 - \frac{V_i}{V_s} \tag{2.50}$$

In the case of non-symmetric activity coefficients one component must take the role of the solvent. The regular solution model is suitable only for non-polar mixtures of molecules that are not too different in size, the activity coefficients always exceed unity, and the excess enthalpy is always positive. Nevertheless, even in the absence of any other information the predictive model is quite useful in certain cases. A compilation of solubility parameters is given by Barton (1991).

2.2.2 Solvatochromic Models

Pure substances in thermodynamics are defined by their acentric factor, critical pressure or temperature, etc. Properties like polarity, acidity, basicity, etc. are usually considered only indirect in that respect. However, in chemistry, a solvent is often characterised by the latter terms. At a molecular scale a solvation process will take place in several stages, although only the overall process is measurable. First, a cavity must be created in the solvent to accomodate the solute. Donor–acceptor bounds between the solute and solvent will develop and dipol orientations of a polar solvent will be induced if the solute is an ion. Once this new aggregate is formed the solvated solute may further interact with its surrounding (hydrogen bonding etc.). The Gibbs free energy change for the process of dissolution of a solute i from phase α to β is zero at equilibrium:

$$0 = \Delta\mu_{s,i}^{\infty} = \frac{\mu_i^{\infty,\alpha} - \mu_i^{\infty,\beta}}{RT} + \ln\frac{x_i^{\alpha}}{x_i^{\beta}} + \ln\frac{\gamma_i^{\infty,\alpha}}{\gamma_i^{\infty,\beta}} \tag{2.51}$$

At infinite dilution the activity coefficients approach unity and with the definition of the distribution coefficient:

$$K_i^{\alpha/\beta} = \frac{x_i^{\alpha}}{x_i^{\beta}}\frac{\gamma_i^{\alpha}}{\gamma_i^{\beta}} \tag{2.31}$$

Equ. (2.51) then reads:

$$\lim_{x\to 0} \ln K_i^{\alpha/\beta} = \ln K_i^{\infty,\alpha,\beta} = \frac{\mu_i^{\infty,\beta} - \mu_i^{\infty,\alpha}}{RT} \tag{2.52}$$

The standard molar Gibbs free energy of solvation, $\Delta\mu_{s,i}^{\infty}$, can be derived from pure component data using spectroscopic information for determining solvatochromic parameters. A generalised equation for $\Delta\mu_{s,i}^{\infty}$ with a linear dependence on solvatochromic parameters is:

$$\Delta\mu_{s,i}^{\infty} = C_0 + C_1\delta_i^2 + C_2\pi_i^* + C_3\alpha_i + C_4\beta_i \tag{2.53}$$

Here δ_i^2 is the cohesive energy density of the solvent which is the square of the Hildebrand solubility parameter; α_i and β_i characterise respectively activity and basicity ability to form hydrogen bonding interactions; π_i^* defines the polarity or polarisability of the solvent.

The estimation of solvatochromic parameters is based on the empirical rule that absorption signals in spectrometry are shifted due to solvent interactions. The estimation of solvatochronic parameters is on an empirical basis which says that the position, ξ, of different solutes (indicators) in spectra of a solvent is a linear function:

$$\xi = C_0 + C_1\,\pi_i^* + C_2\,\alpha_i + C_3\,\beta_i \tag{2.54}$$

$C_0,\ldots,C_3$ are properties of the indicator and π^*, α_i and β_i are properties of the solvent. With different analytical methods (NMR, UV-VIS etc.) and different indicators the solvatochronic parameters of a solvent can be estimated. These linear relationships do not only correlate spectroscopic positions of indicators in different solvents. They can also be used to correlate the influence of the solvent on reaction equilibria and kinetics. In principle it considers differences in the Gibbs' free energy resulting from a change of the electron configurations in a molecule.

There exists a number of such linear solvatochromic scales. One of the most widely used is that of Kamlet and Taft which is the basis of the LSER[1] [Kamlet et al. (1981, 1983)]. The application of LSER can be found in different fields, e.g. distribution coefficients (retention times) in chromatography [Carr (1993)], solubility of *HCl* in different solvents [Luehrs & Godbole (1994)] and liquid-liquid distribution coefficients [Kamlet et al. (1988)]. The Nernst distribution according to Kamlet is:

[1] LSER: linear solvation energy relationships

$$\log D_{ij}^{\infty} = C_0 + C_1\, V_i\ + C_2\ \partial_i + C_3\, \pi^* + C_2\, \alpha_i + C_2\, \beta_i \tag{2.55}$$

Here the cohesive energy density in equ. (2.54) is replaced by the molar volume V_i of the solute (as a measure of the size of the cavity to accomodate the solute i in the solvent) and ∂_i is an empirical parameter which takes also account for polarisability π^*. Meyer and Maurer (1995) used this equation for 30 systems (371 substances, 947 experimental distribution coefficients) to evaluate generalised solvent (j) independent parameters (k = 0 .... 5) as to:

$$C_k = C_0^{(k)} + C_1^{(k)}\, V_j + C_1^{(k)}\, \partial_j + C_1^{(k)}\, \pi_j + C_1^{(k)}\, \alpha_j + C_1^{(k)}\, \beta_j \tag{2.56}$$

This equation is quite accurate in comparison with group contributing methods [Meyer & Maurer (1993)] or other predictive LSER methods [Marcus (1991)]. For compounds where the solvatochromatic parameters are known, the mean absolute error in log D_{ij} is about 0.16. It is usually less than 0.3 if solvatochromatic parameters of the solute and solvent have to be estimated according to empirical rules [Kamlet, et al. (1988)]. In contrast to the prediction of gas–liquid distribution coefficients, which is usually easier, the LSER-method allows a robust estimation of liquid-liquid distribution coefficients. However, such equations always involve empirical terms, despite of the physico-chemistry founded thermodynamic models due to the fundamental character of the solvatochromic scales.

As an alternative to the LSER method the MOSCED[2] has been developed in order to correlate and predict activity coefficients at infinity dilution [Thomas & Eckert (1984), Howell et al. (1989)]. An advantage especially with liquid-liquid systems is that experimental distribution coefficients and limiting activity coefficients are known at ambient temperatures. Thus it is very convenient to develop correlations on the basis of solvatochromic parameters. A further development of the MOSCED method is the SPACE[3] concept [Hait et al. (1993)]:

$$\gamma_i^{\infty} = \frac{V_i}{RT}\left((\lambda_s - \lambda_i)^2 + (\tau_s - \tau_i)^2 + (\tilde{\alpha}_s - \tilde{\alpha}_i)(\tilde{\beta}_s - \tilde{\beta}_{ii})\right) + \ln \gamma_i^{\infty,\text{F-H}} \tag{2.57}$$

with the modified Flory–Huggins term similar to equ. (2.50):

$$\ln \gamma_i^{\infty,\text{F-H}} = \ln\left(\frac{V_i}{V_s}\right)^a + 1 - \left(\frac{V}{V_s}\right)^a \tag{2.58}$$

(λ_s - λ_i) considers dispersive interaction. λ_i is then an empirical function of the refractive index. Polar interactions are in (τ_s - τ_i). τ_i is linearly correlated with the solvatochromic parameter π^*. $(\tilde{\alpha}_s - \tilde{\alpha}_i)(\tilde{\beta}_s - \tilde{\beta}_i)$ subsums *H*-bridge contributions. $\tilde{\alpha}_i$ and $\tilde{\beta}_i$ are linear correlated with the solvatochromic parameters α_i and β_i. For the solvatochromic parameters of the solvent (s) the spectroscopically derived values after Kamlet and Taft are used, for the solute

[2] MOSCED: modified separation of cohesive energy density model

[3] SPACE: solvatochromic parameters for activity coefficient estimation

(*i*) data from gas chromatography [Li et al. (1993)] are used. The SPACE method is an alternative to the predictive group contribution models and can currently be used only for non-aqueous systems, e.g. with distillation problems. The errors are usually within 10% considering relatively ideal mixtures.

2.2.3 Margules Equation

As already mentioned, in a binary mixture the activity coefficients are related to the Gibbs excess energy as to:

$$RT\ln\gamma_1 = \frac{\partial G^E}{\partial n_1} \qquad RT\ln\gamma_2 = \frac{\partial G^E}{\partial n_2} \tag{2.59}$$

However, the Gibbs excess energy must become zero for a pure component. Thus any model must tend to zero when any of the mole fraction equals unity. The simplest model is as follows where A is a measure of the non-ideality:

$$\frac{g^E}{RT} = Ax_1x_2 \qquad \text{or} \qquad \frac{G^E}{RT} = A\frac{n_1n_2}{n_1+n_2} \tag{2.60}$$

The partial derivatives of the Gibbs excess energy yield the activity coefficients for this model:

$$\ln\gamma_1 = Ax_1x_2$$

$$\ln\gamma_2 = Ax_1^2 \tag{2.61}$$

Both functions are depicted in Fig. 2.1 and strong non-ideality can be seen for $A > 2$ (bold line).

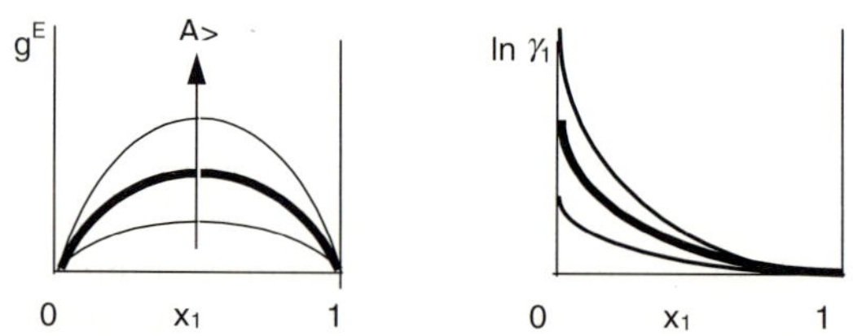

Fig. 2.1: A simple g^E function and its activity coefficients

This simple model can be easily extended to multicomponent mixtures and for a ternary we obtain:

$$\frac{g^E}{RT} = A_1 x_1 x_2 + A_2 x_2 x_3 + A_3 x_3 x_1 \tag{2.62}$$

or

$$\frac{G^E}{RT} = A_1 \frac{n_1 n_2}{n_1 + n_2 + n_3} + A_2 \frac{n_2 n_3}{n_1 + n_2 + n_3} + A_3 \frac{n_3 n_1}{n_1 + n_2 + n_3} \tag{2.63}$$

As can be seen, this model contains only one binary interaction parameter A_i for each binary system. In order to account for interactions of molecules of different size, the introduction of a second binary parameter leads to the Margules equation:

$$\frac{g^E}{RT} = x_1 x_2 \left(A + B(x_1 - x_2)\right) \tag{2.64}$$

A positive value of B skews the Gibbs excess energy to the right and left for a negative value, allowing the activity coefficients to behave non-symmetrically (see Fig. 2.2).

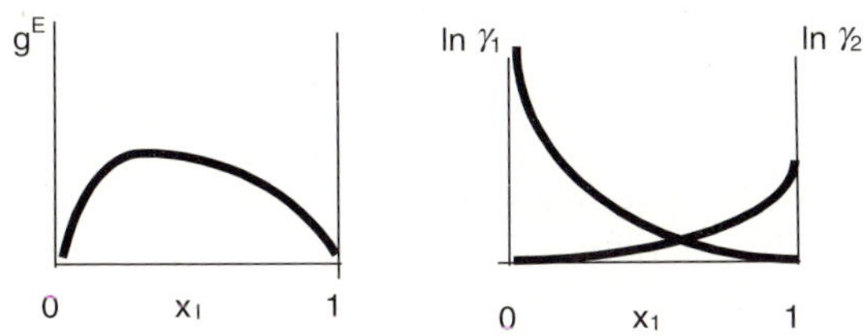

Fig. 2.2: The asymmetric Margules g^E model with its activity coefficients

Other g^E models (van Laar, Wilson, NRTL[4], etc.) are more complicated and require two or more parameters for each binary. Among these the van Laar and Wilson models are for moderate non-ideal mixtures and thus cannot handle liquid-liquid phase splitting [van Ness & Abbott (1997)].

2.2.4 NRTL Model

More recent equations like the NRTL model [Renon & Prausnitz (1968)] are based on a semi-empirical physical model, called the two-liquid theory of Scott (1956) combined with the local composition concept. Both can be used for mixtures in which phase splitting can occur. The local composition on a molecular level differs from the macroscopic composition. The molecules do not mix randomly because of interactions with their surroundings (polarity

[4] NRTL: non random two liquids

effects, hydrogen bonding, etc.). However, the local mole fraction cannot be measured easily but must be related to the overall composition.

A Boltzmann factor from statistical thermodynamics relates the local mole fractions, x_{ij}, to the overall ones, x_j:

$$\frac{x_{ji}}{x_{ii}} = \left(\frac{x_j}{x_i}\right) G_{ji} \tag{2.65}$$

and

$$G_{ji} = \exp(-\alpha_{ij}\, \tau_{ji}) \tag{2.66}$$

with

$$RT\tau_{ji} = (g_{ji} - g_{ii}) \tag{2.67}$$

The three model parameters are the residual Gibbs energies g_{ij} and g_{ii} and an empirical "non-randomness" parameter, α_{ij}, which characterises a binary system. Both g_{ij} and α_{ij} are inherently symmetric (g_{ij} = g_{ji} and α_{ij} = α_{ji}), thus τ_{ii} = 0 and g_{ii} = 0. The general expression for a multicomponent mixture is then:

$$\ln \gamma_i = \frac{\sum_j x_j \tau_{ji} G_{ji}}{\sum_k x_k G_{ki}} + \sum_j \left\{ \frac{x_j G_{ji}}{\sum_k G_{kj} x_k} \left[\tau_{ij} - \frac{\sum_m x_m \tau_{mj} G_{mj}}{\sum_k G_{kj} x_j} \right] \right\} \tag{2.68}$$

The "non-randomness" parameter can be identified by the inverse number of the nearest neighbours of a molecule and its local environment and is usually in the range from 0.1 to 0.3. These limits have been found empirically and in liquid-liquid extraction a value of 0.2 is often assigned [Sorensen et al. (1979)]. The binary parameters of the NRTL equation can be determined by fitting activity coefficients obtained from VLE data and τ_{ij} is often found to vary linearly with temperature.

2.2.5 UNIQUAC Model

Abrahams and Prausnitz (1975) have combined the local composition concept with the quasi-chemical lattice model of Guggenheim (1952) to a truly two parameter model – UNIQUAC[5]. Here the Gibbs excess energy results from two additive parts, a combinatorial term, g^C, accounting for molecular size and shape differences, and a residual term, g^R (not a residual property), accounting for the molecular interaction energies:

$$\frac{g^C}{RT} + \frac{g^R}{RT} = \frac{g^E}{RT} \tag{2.69}$$

The function g^C contains only pure species parameters:

[5]UNIQUAC: universal quasi chemical theory

$$\frac{g^C}{RT} = \sum_i x_i \, ln \frac{\Phi_i}{x_i} + 5 \sum_i q_i x_i \, ln \frac{\theta_i}{\Phi_i} \tag{2.70}$$

and g^R contains two binary parameters u_{ij} and u_{ji} which are symmetric ($u_{ij} = u_{ji}$ and $u_{ii} \neq u_{jj}$):

$$\frac{g^R}{RT} = -\sum_i q_i x_i \ln\left(\sum_j \theta_j \tau_{ji}\right) \tag{2.71}$$

where

$$\Phi_i \equiv \frac{x_i r_i}{\sum_j x_j r_j} \tag{2.72}$$

and

$$\theta_i \equiv \frac{x_i q_i}{\sum_j x_j q_j} \tag{2.73}$$

The temperature dependence on τ_{ij} (note $\tau_{ij} \neq \tau_{ji}$ and $\tau_{ii} = \tau_{jj} = 1$) is:

$$\tau_{ij} = \exp - \left(u_{ji} - u_{ii}\right) / \, RT \tag{2.74}$$

In this model the parameters Φ_i and θ_i are the average segment and area fractions, respectively, and they are related to the pure species parameter r_i (relative van der Waals molecular volume) and q_i (relative van der Waals surface). However, for molecules of equal size and shape is $x_i = \theta_i = \Phi$ and thus $g^C = 0$ and for equal energy interactions $\Delta u_{ii} = \Delta u_{jj} = 0$ and thus $g^R = 0$. Hence the expression for the activity coefficient is also combinatorial:

$$\ln \gamma_i = \ln \gamma_i^C + \ln \gamma_i^R \tag{2.75}$$

$$\ln \gamma_i^C = 1 - J_i + \ln J_i - 5q_i \left(1 - \frac{J_i}{L_i} + \ln \frac{J_i}{L_i}\right) \tag{2.76}$$

$$\ln \gamma_i^R = q_i \left(1 - \ln s_i - \sum_j \theta_j \frac{\tau_{ij}}{s_j}\right) \tag{2.77}$$

with

$$J_i = \frac{r_i}{\sum_j r_j x_j} \tag{2.78}$$

$$L_i = \frac{q_i}{\sum_j q_j x_j} \tag{2.79}$$

$$s_i = \sum_k \theta_k \tau_{ki} \tag{2.80}$$

Likewise for the NRTL model the binary parameters can be obtained from VLE data (solubilities, aceotropic data, activity coefficients at infinite dilution) or from LLE data [Gmehling et al. (1977–1999), Sorensen & Arlt (1979–1980)]. Usually the latter ones are more precise and VLE data should only be used for qualitative calculations in the extraction field. A special table of parameters for LLE calculations is given by Magnussen et al. (1981).

Both, the UNIQUAC and the NRTL model have problems predicting phase equilibria in systems containing associating solutes [Maurer 1996]. Alcohols, phenols, amines, carboxylic acids, water, etc. have the ability for hydrogen bonding due to donor–acceptor interactions. In the vapour phase these associates are mainly dimeric, but in the liquid phase higher aggregates are likely. In the vapour phase this dimerisation is considered with the "chemical" theory [Prigogine & Defay (1954)] where the dimerisation constants come from spectroscopic and densiometric investigations.

Association g^E models have been reviewed by Marsh & Kohler (1985) and more recently by Hasse (1996). In contrast to the works of Stokes (1977), Kohler (1972), Nagata & Miyamoto (1993) the model of the Eckert group [Karachewski et al. (1991)] is very simple and the association parameters are derived from NMR spectroscopy. This approach is vital, otherwise an evaluation of these parameters only from thermodynamic data sets reduces these association equilibrium constants to a pure fit parameter. However, the spectroscopic information is only necessary for the combinatorial term (true size of the associative van der Waals volume and surface) and not the binary residual interaction term [Asprion et al. (1998)]. Using the alternative approach it is possible to divide the van der Waals surface into zones, which according to their polarity or *H*-bridging ability exhibit different interaction energies. Thus hydrogen bonding effects, dipole–dipole interactions and isomeric orientation aspects can be better handled with DISQUAC [Kehiaian et al. (1978)] or with GEQUAC [Egner et al. (1997)]. As can be seen, active research is in that field and further improvements might be expected in the near future.

2.2.6 Group Contribution Methods

In the absence of any experimental information, the regular solution theory and the concepts of LSER, MOSCED, SPACE, etc. offer the possibility to predict the thermodynamic behaviour of information from pure substance data. Another approach is to use group contribution methods since an infinite number of chemical compounds exist but only a limited number of functional groups. It is convenient to estimate functional group parameters from existing data and then predict the behaviour of new compounds. With the concepts mentioned above, which rely on the regular solution theory, exists a limitation to rather non-polar compounds. Polar ones can be better handled with modern group-contribution concepts like ASOG[6] [Kojima & Tochigi (1979)] or UNIFAC[7] [Fredenslund et al. (1975)]. However, higher molecular weight compounds with more than one functional group will be badly represented or the inclination of the conodes will be wrongly predicted. Nevertheless, these methods give quite a good qualitative view on the system and are handy tools for feasibility studies and a first process lay-out.

The UNIFAC concept assembles molecular properties from groups and like the UNIQUAC concept comprises two parts:

$$\ln \gamma_i = \ln \gamma_i^C + \ln \gamma_i^R \tag{2.81}$$

The combinatorial part is analogous to the UNIQUAC calculation when the van der Waals parameters are calculated as follows:

$$r_i = \sum_k \nu_k^{(i)} R_k \tag{2.82}$$

$$q_i = \sum_k \nu_k^{(i)} Q_k \tag{2.83}$$

The summation includes the number of structural groups $\nu_k^{(i)}$ of type k in solute i. The group surface parameters, R_K and Q_K, can be found in tables [Fredenslund et al. (1977)]. The residual energy term arises from contributions of group activities as:

$$\ln \gamma_i^R = \sum \nu_k^{(i)} (\ln \Gamma_k - \ln \Gamma_k^{(i)}) \tag{2.84}$$

The concentration dependence of the mixture Γ_k or of the pure solute $\Gamma_k^{(i)}$ is as in the UNIQUAC model:

$$ln\Gamma_k = Q_k \left[1 - ln\left(\sum_m \Theta_m \Psi_{mk} \right) - \sum_m \frac{\Theta_m \Psi_{km}}{\sum_m \Theta_n \Psi_{nm}} \right] \tag{2.85}$$

with the surface fraction Θ_m of group m in the mixture X_m:

$$\Theta_m = \frac{Q_m X_m}{\sum_n Q_n X_n} \tag{2.86}$$

$$X_m = \frac{\sum_j \nu_m^{(j)} x_j}{\sum_j \sum_n \nu_n^{(j)} x_j} \tag{2.87}$$

and the temperature dependence of the interaction parameter ψ_{nm}:

$$\Psi_{nm} = \exp\left(-\frac{a_{nm}}{T} \right) \tag{2.88}$$

---

[6] ASOG: analytical solution of groups

[7] UNIFAC: UNIQUAC functional group activity coefficients

where a_{nm} is a binary parameter between group n and m. Each combination of groups requires the knowledge of two binary parameters (a_{nm} and a_{mn}) whereas by convention CH_3, CH_2, CH, C are one group. However, conformational information (*cis/trans*, *iso* etc.) can thus not be related but nevertheless an appropriate data bank exists, which also allows calculation of phase equilibria with compounds including sulfolane, furvoral, tetrahydrofurvoryl alcohol, etc. [Magnussen et al. (1981), Gmehling et al. (1993)]. The available data banks are, e.g. DECHEMA (Internet), FIZ Karlsruhe (STN), and DECHEMA Chemistry Data Series.

The latest developments result in two modifications, modUNIFAC (Dortmund) by Weidlich and Gmehling (1987) and modUNIFAC (Lyngby) by Larsen (1986) which improve the prediction of VLE. However, the LLE are still a problem. Thus far, a new type of predictive model has been recently developed called COSMO-RS (http://www.cosmologic.de). Thermodynamic parameters, even of complex molecules, are derived by a molecular dynamics simulation based on the density functional theory. Except for electrolytes, it is capable of treating almost the entire equilibrium thermodynamics at a similar accuracy as UNIFAC. The model needs eight general COSMO-RS inherent parameters, a single radius and one dispersion constant per chemical element. Due to its molecular dynamic basis it properly predicts differences in the properties of enantiomers, isomers, etc. [Klampt et al. (1998), Clausen et al. (1999)]; so far it has only been intensively tested for VLE. Another interesting approach is to use a group contribution equation of state [Horst et al. (2000)] which again is only tested for VLE.

2.3 Electrolyte Systems

Electrolyte systems play a vital part in many solvent extraction processes, especially when ionic solutes are extracted by means of liquid ion exchangers. Anions, cations and even non-dissociated compounds will be extracted and the extractability mainly depends on the dissociation equilibria of complexes, salts and acids. However, the chemical equilibrium has to be considered in a similar manner as discussed above with associating solutes.Thus, the introduction of ionic interactions brings a new quality to be considered into the system.

As mentioned initially, the thermodynamic equilibrium comprises thermal, physical, mechanical and reaction equilibrium in such reacting systems.

In regard to the reaction equilibrium the law of mass action yields for a single reaction R:

$$K_R = exp\left(\frac{-\sum_{i=1}^{n} \nu_{i,R}\, \mu_i^0}{RT} \right) \tag{2.89}$$

or in molal units:

$$K_R = \prod_{i \neq w} (m_i \gamma_i^{(m)})^{\nu_{i,R}} \tag{2.90}$$

Additionally, as a summation condition, electroneutrality is required in electrolyte systems:

$$\sum_{i=1}^{n} n_i \cdot z_i = 0 \tag{2.91}$$

which means the charges, z, of cations and anions must be balanced. However, in ionic solutions the reference state is usually at a molal scale ($m^{\oplus}$ = 1 mol/kg water) and thus also the activity coefficient

$$a_i = \gamma_i^{(m)} \frac{m_i}{m^{\oplus}} \tag{2.92}$$

Henceforth, the superscript (m) for molal will be omitted and a conversion from molal to molar (c) or mole fraction (x) is given in the appendix A.

In a multicomponent–multiphase system one has to consider all reactions in any phase, the appropriate phase equilibria (Raoult, Henry, Nernst), the mass balances and, if ions are involved, the charge balance as well. In order to better understand the previous equations, let us consider a simple example, the water–chlorine system. The reactions involved are:

$$H_2O^L \leftrightarrow H_2O^G \tag{2.93}$$

$$Cl_2^L \leftrightarrow Cl_2^G \tag{2.94}$$

$$Cl_2^L + H_2O^L \leftrightarrow H^+ + Cl^- + HClO^L \tag{2.95}$$

$$HClO^L \leftrightarrow H^+ + ClO^- \tag{2.96}$$

$$H_2O^L \leftrightarrow H^+ + OH^- \tag{2.97}$$

For a predictive treatment of the phase distribution and compositions the following equilibria have to be taken into account:

$$K_{H_2O}^G = \frac{p\, y_W\, \varphi_W^G}{p_W^0\, a_w} \tag{2.98}$$

$$K_{Cl_2}^G = \frac{p\, y_{Cl_2}\, \varphi_{Cl_2}^G}{H_{Cl_2}\, m_{Cl_2}\, \gamma_{Cl_2}} \tag{2.99}$$

$$K_{Cl_2}^L = \frac{(\gamma_{H^+}\, m_{H^+})(\gamma_{Cl^-}\, m_{Cl^-})(\gamma_{HClO}\, m_{HClO})}{(\gamma_{Cl_2}\, m_{Cl_2})\, a_w} \tag{2.100}$$

$$K_{HClO}^L = \frac{(\gamma_{H^+}\, m_{H^+})(\gamma_{ClO^-}\, m_{ClO^-})}{(\gamma_{HClO}\, m_{HClO})} \tag{2.101}$$

$$K_{H_2O}^L = \frac{(\gamma_{H^+}\, m_{H^+})(\gamma_{OH^-}\, m_{OH^-})}{a_w} \tag{2.102}$$

The electroneutrality condition in that system delivers:

$$m_{H^+} = m_{OH^-} + m_{Cl^-} + m_{ClO^-} \tag{2.103}$$

Additionally, in the liquid phase the following overall mass balances (e.g. for pauschal chlorine and hydrogen) must hold:

$$\overline{m}_{Cl} = 2\, m_{Cl_2} + m_{HClO} + m_{Cl^-} + m_{ClO^-} \tag{2.104}$$

$$\overline{m}_{H} = 2\, m_{H_2O} + m_{HClO} + m_{H^+} + m_{OH^-} \tag{2.105}$$

and finally the Dalton's law is valid:

$$p = p_w \cdot \varphi_w^G + p_{Cl_2} \cdot \varphi_{Cl_2}^G \tag{2.106}$$

This set of nine equations serves for the nine species (n = 9) in this two phase system with three reaction equilibria (k = 3).

An extended review on thermodynamics of electrolyte systems is given in the textbooks of Horvath (1985) or Zermaitis et al. (1986). From a historical point of view, the first consistent thermodynamic approach to electrolyte solutions was made by Debye and Hückel (1923) which was then further improved by Pitzer (1973) to tackle high concentrated solutions. Multicomponent behaviour is nowadays modelled on the basis of the local composition concept incorporating the Coulomb interactions, due to the Pitzer concept.

In electrolyte solutions one will observe following interactions as are:

- molecule–molecule,
- molecule–ion and
- ion–ion.

The short-range forces between neutral molecules diminish according to $1/R^6$ and they are well described with the local composition concept and have been applied in several models (e.g. NRTL, UNIQUAC). For the long range Coulomb forces ($1/R$ decay) between ions and their environment Debye and Hückel (1923) developed the following equation based on a molal scale which is still up to now a cornerstone for more advanced models:

$$\log \gamma_\pm = -A_m \left| z_+ \, z_- \right| \sqrt{I} \tag{2.107}$$

where I is the ionic strength and z the charge of the ions:

$$I = \frac{1}{2} \sum_i^n m_i \, z_i^2 \tag{2.108}$$

A_m is referred to as the Debye–Hückel constant and is temperature dependent. It appears in all recent electrolyte activity coefficient models:

$$A_m = \frac{1}{3} \left(2\pi \, N_A \, \rho \right)^{0,5} \left(\frac{e^2}{4\pi \, \varepsilon_0 \, \varepsilon_r \, k \, T} \right)^{1,5} \tag{2.109}$$

Here ε_0 is the electric field constant, ε_r the electric constant, e the elementary charge, N_A the Loschmidt number, k the Boltzmann constant, and ρ the density of solvent. Due to the assumptions and simplifications this limiting law is only valid at very diluted solutions of an ionic strength of 0.001 molal or less. Especially with higher concentrated electrolytes the short range ion–ion interactions cannot longer be neglected.

A summary of electrolyte activity coefficient models, examples and data are given by Zermaitis et al. (1990). Here a milestone is the Pitzer model [Pitzer (1973)], which is a virial type equation for the excess Gibbs energy and is an improvement upon an earlier model proposed by Guggenheim (1935, 1969). The Pitzer model analyses 'hard-core'effects of the Debye–Hückel theory and can thus be used for aqueous electrolyte systems up to 6 molal ionic strength. In its modified form [Edwards et al. (1978)] it reads for water as solvent:

$$\frac{1000\, G^E}{R\, T\, n_w\, M_w} = f_1(I) + \sum_{i \neq w} \sum_{j \neq w} m_i\, m_j\, \lambda_{ij}(I) + \sum_{i \neq w} \sum_{j \neq w} \sum_{k \neq w} m_i\, m_j\, m_k\, \tau_{ijk} \tag{2.110}$$

Here $f_1\ (I)$ is a Debye–Hückel term which was modified by Pitzer with b estimated to be 1.2 (kg/mol)$^{0.5}$:

$$f_1(I) = -A_m \cdot \frac{4 \cdot I}{b} \cdot \ln(1 + b\sqrt{I}) \tag{2.111}$$

The binary interaction parameters λ_{ij} are according to Pitzer dependent on the ionic strength, whereas the ternary τ_{ijk} are not:

$$\lambda_{ij} = \beta_{ij}^{(0)} + \frac{2\beta_{ij}^{(1)}}{\alpha^2 \cdot I} \cdot \left(1 - \left(1 + \alpha \cdot \sqrt{I}\right) \cdot \exp\left(-\alpha \cdot \sqrt{I}\right)\right) \tag{2.112}$$

Here $\beta_{ij}(0)$ and $\beta_{ij}(1)$ are the binary, temperature dependent interaction parameters. For 1:1 and 1:2 electrolytes α = 2 (kg/mol)$^{0.5}$ according to Pitzer. The binary and ternary parameters are symmetric:

$$\beta_{ij}^{(0)} = \beta_{ji}^{(0)} \tag{2.113}$$

$$\beta_{ij}^{(1)} = \beta_{ji}^{(1)} \tag{2.114}$$

$$\tau_{ijk} = \tau_{ikj} = \tau_{jki} = \tau_{jik} = \tau_{kij} = \tau_{kji} \tag{2.115}$$

The activity coefficient of a solute ($i \neq w$) is obtained from the Gibbs excess energy as usual:

$$\ln \gamma_i = \frac{1}{RT} \left(\frac{\partial G^E}{\partial n_i} \right)_{T, p, n_{j \neq i}} \tag{2.116}$$

thus:

$$\ln\gamma_i = -A_w\, z_i^2\left[\frac{\sqrt{I}}{1+b\sqrt{I}} + \frac{2}{b}\ln\left(1+b\sqrt{I}\right)\right] + 2\sum_{j\neq w} m_j\left[\beta_{ij}^{(0)} + \beta_{ij}^{(1)}\, f_2(I)\right]$$
$$- f_3(I)\, z_i^2 \sum_{j\neq w}\sum_{k\neq w} m_j\, m_k\, \beta_{jk}^{(1)} + 3\sum_{j\neq w}\sum_{k\neq w} m_j\, m_k\, \tau_{ijk} \tag{2.117}$$

The terms $f_2(I)$ and $f_3(I)$ read as:

$$f_2(I) = \frac{2}{\alpha^2\, I^2}\left(1-\left(1+\alpha\sqrt{I}\right)\exp\left(-\alpha\sqrt{I}\right)\right) \tag{2.118}$$

$$f_3(I) = \frac{1}{\alpha^2\, I^2}\left[1-\left(1+\alpha\sqrt{I}+\frac{1}{2}\alpha^2\, I\right)\exp\left(-\alpha\sqrt{I}\right)\right] \tag{2.119}$$

The activity of water can be derived with the help of the Gibbs–Duhem equation to:

$$\ln a_w = \frac{M_w}{1000}\left[\frac{1000\, G^E}{R\, T\, n_w\, M_w} - \sum_{i\neq w} m_i\, \ln\gamma_i - \sum_{i\neq w} m_i\right] \tag{2.120}$$

and inserting equs. (2.110) and (2.117) yields

$$\ln a_w = \frac{M_w}{1000}\left[\begin{array}{l} 2\, A_m\, \dfrac{I^{1.5}}{1+b\sqrt{I}} - \sum\sum m_i\, m_j\left(\beta_{ij}^{(0)} + \beta_{ij}^{(1)}\, \exp-\left(\alpha\sqrt{I}\right)\right) \\ -2\sum_{i\neq w}\sum_{j\neq w}\sum_{k\neq w} m_i\, m_j\, m_k\, \tau_{ijk} - \sum_{i\neq w} m_i \end{array}\right] \tag{2.121}$$

In order to tackle weak electrolyte systems with molecular solutes, the second and third virial coefficients in the basic Pitzer equation for molecule–ion and molecule–molecule interactions must be considered. However, binary and ternary parameters between species of equal charge are usually set to zero.

As an alternative to the modified Pitzer model the NRTL model is likewise used. It is based on two fundamental assumptions:

- The local composition n of cations (anions) around a central cation (anion) is zero. The same is valid for anions around a central anion.
- The distribution of anions and cations around a central molecule is such that the net local charge is zero.

The model reduces to the Debye–Hückel formula at infinitely diluted electrolyte solution; in the absence of electrolytes to the NRTL model [Renon & Prausnitz (1968)]. It has been also extended to model mixed solvent electrolyte systems and can be used even for non-aqueous solvents (MNRTL) [Mock et al. (1986)]. Active research is being conducted in the field of electrolyte thermodynamics [Maurer (1983, 1996)] thus a broad range of thermodynamically consistent activity coefficient models will be available. A promising modification of UNIFAC is LIFAC [Li et al. (1994), Polka et al. (1994)] which allows the prediction of VLE in electrolyte systems. Here the long range Coulomb forces are considered in terms of the Debye–Hückel theory and the middle range charge effects with the Pitzer theory. Short range interactions in accordance with the Gibbs excess energy are calculated using the UNIFAC model.

Commercially available flowsheeting simulators (e.g. ASPENPLUS®, CHEMCAD®, etc.) give access to the electrolyte activity models with an appropriate parameter data base. For solutions of common industrial interest, special inserts can be loaded (e.g. with ASPEN BRINE (H_2O, CO_2, H_2S, $NaCl$), CAUST (H_2O, $NaCl$, Na_2SO_4, $NaOH$), ...), where all the necessary reaction equilibria parameters are included.

2.4 Ion Exchange Equilibria in Liquid–Liquid Systems

As has been described in the first chapter, the use of liquid ion exchangers converts aqueous ionic and non-ionic solutes into reextractable organic complexes. According to the type of ion exchanger used, the reactive phase equilibrium is strongly shifted by either temperature, ionic strength, acidity or excess of extractable counter ions. The description of the non-ideal behaviour of the organic and aqueous species can be achieved by using an activity coefficient model as presented above. In the following analysis we will use the Pitzer model for the aqueous species and the Hildebrand–Scott model for the organic ones.

The structure of the solute–ion exchanger complex determines the stoichiometry of the overall extraction reaction. It should be independently developed from a different data source like the thermodynamic data, otherwise it is reduced to a pure fit parameter. This is discussed with cation and anion exchange as an example, applying spectrometric methods and slope analysis to reveal the complex stoichiometry.

2.4.1 Metal/Cation Exchange System

The cation exchanger di(2-ethylhexyl)phosphoric acid (*D2EHPA*) is a well known extractant, usually forms dimers in aliphatic diluents and is mainly monomeric in aromatic ones [Kolarik (1982)]. The overall extraction of zinc with dimeric *D2EHPA* (R_2H_2) is as follows:

$$aZn^{2+} + b\overline{R_2H_2} \leftrightarrow \overline{Zn_aR_{2\cdot a}(RH)_{2\cdot b-2\cdot a}} + 2H^+ \tag{2.122}$$

At higher concentrations polynuclear *Zn-D2EHPA* complexes are known to occur [Kolarik & Grimm (1976), Grimm & Kolarik (1974a,b), Danesi & Vandegrift (1981), Koncar et al. (1988)]. These complexes can have up to eight zinc-ions, this means the nucleation number a is in the range from 1 to 8

[Bart et al. (1994)]. However, at lower zinc concentrations, there are only mononuclear zinc complexes to consider:

$$Zn^{2+} + b\overline{R_2H_2} \leftrightarrow \overline{ZnR_2(RH)_{2\cdot b-2}} + 2H^+ \tag{2.123}$$

The law of mass action on the basis of concentrations is then:

$$K_{1,2b} = \frac{[\overline{ZnR_2(RH)_{2\cdot b-2}}][H^+]^2}{[Zn^{2+}][\overline{R_2H_2}]^b} = D_{Zn}\frac{[H^+]^2}{[\overline{R_2H_2}]^b} \tag{2.124}$$

where D_{Zn} denotes the distribution coefficient of zinc between organic and water phase, as defined in Chapter 1. As has been discussed there, a logarithmic formulation yields:

$$\log D_{Zn} = b\log\left([\overline{R_2H_2}]\right) + 2\text{pH} + \log K_{1,2b} \tag{2.125}$$

At low solute concentrations the total *D2EHPA* concentration can be approximated by the initial concentration:

$$[\overline{R_2H_2}] = [\overline{R_2H_2}]_0 - b[\overline{ZnR_2(RH)_{2\cdot b-2}}] \cong [\overline{R_2H_2}]_0 \tag{2.126}$$

Additionally, at a low initial pH value the concentration of H^+ during the extraction will thus not change significantly, keeping the pH value almost constant:

$$[H^+] = [H^+]_0 + 2[\overline{ZnR_2(RH)_{2\cdot b-2}}] \cong [H^+]_0 \tag{2.127}$$

As a result, the distribution coefficient D_{Zn} is not a function of $[Zn^{2+}]$ but log D_{Zn} is a linear function of $[R_2H_2]$ with the slope b. These conclusions are only valid if only one monomeric zinc complex is present in the organic phase. The logarithm of the distribution coefficient versus the logarithm of the dimeric *D2EHPA* concentration is the basis of the slope analysis. The slope represents the stoichiometric ratio of R_2H_2/Zn of the organic zinc complex and thus b in equ. (2.125) and can be derived from Fig. 2.3. The slope at this low zinc concentration (0.1 mM) is approximately 1.5, which is equivalent to three *D2EHPA* per one zinc molecule in the organic complex:

$$Zn^{2+} + 1.5\overline{R_2H_2} \leftrightarrow \overline{ZnR_2(RH)} + 2H^+ \tag{2.128}$$

However, the amount of co-extracted water bound in the solute–ion exchanger complex has to be considered as well. This can be detected, e.g. with thermally enhanced Karl-Fischer titration. At different temperature levels the physical and the chemical extracted water can be determined [Philipp et al. (1997)]. Fortunately, in the system zinc/*D2EHPA* dissolved in alkanes, the water content is negligible. There is a minor increase of the water content with rising *D2EHPA* concentration, which gives a ratio of one molecule water per 20 molecules ion exchanger. This can be attributed to a solvation effect around the phosphoric acid molecule and no additional water is observed in the *Zn-D2EHPA* complex, as can be verified by FTIR-measurements [Sainz-Diaz et al. (1996)]. This minor amount of water co-extracted is in contrast to e.g. nickel extraction with *D2EHPA*, where even water is extracted in inverted micelles [Neumann et al. (1993), Park & Neumann (1992), Neumann & Park

(1992)]. A typical FTIR spectrum of a *Ni-D2EHPA* complex ([*RH*]= 0.150 M, [*Ni*] = 0.0712 M) is depicted in Fig. 2.4.

The *O–H* bands for water at 3400 cm^{-1} and 1630 cm^{-1} can be clearly seen, which is in contrast to the *Zn-D2EHPA* complex depicted at the same conditions in Fig. 2.5.

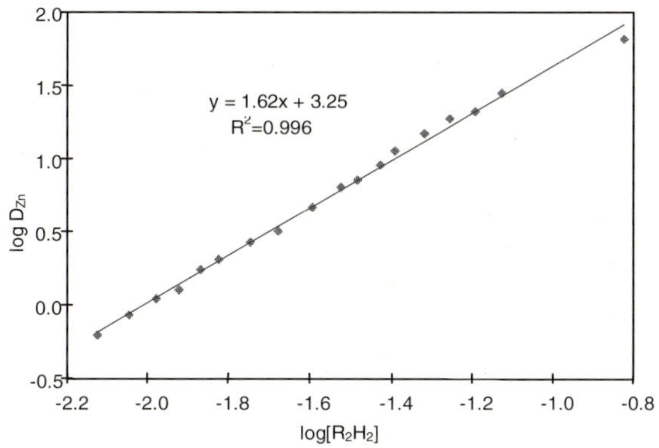

Fig. 2.3: Slope analysis of the *Zn-D2EHPA* complex

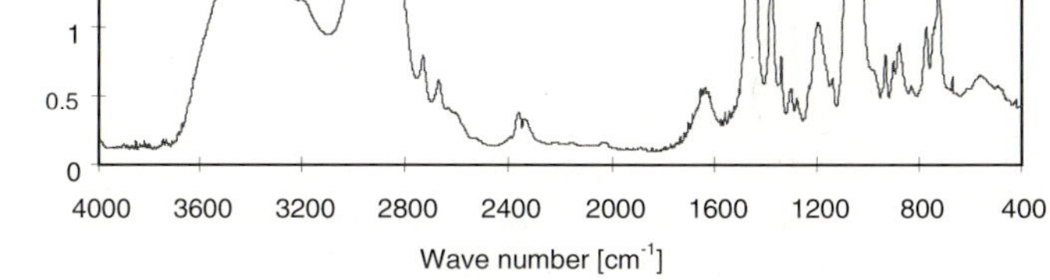

Fig. 2.4: FTIR-spectrum of *Ni-D2EHPA* in n-heptane

The most characteristic IR vibrational bands of *D2EHPA*, in a pure liquid, appear at 1230 cm^{-1} (stretching vibrational band of the *P=O* bond, ν (*P=O*)), and 1034 – 990 cm^{-1} (stretching vibrational band of the *P–O* bond coupled with the *C–O* and *O–H* bonds, ν (*P–O*)) [Sainz-Diaz et al. (1995)]. The band at 1690 cm^{-1} corresponds to the *O–H* bonding vibration [Sato & Nakamura

(1972)]. These bands are more intensive than those corresponding to the 2-ethylhexyl groups (3000 – 2850, 1464 and 1380 cm^{-1}). In n-heptane the $\nu\,(P = O)$ band appears at a slightly higher frequency (1234 cm^{-1}), which can be attributed to the intermolecular interactions between the phosphoric moieties.

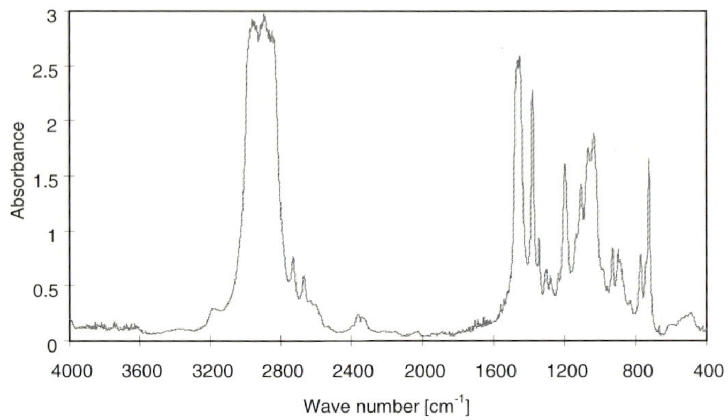

Fig. 2.5: FTIR-spectrum of *Zn-D2EHPA* in n-heptane

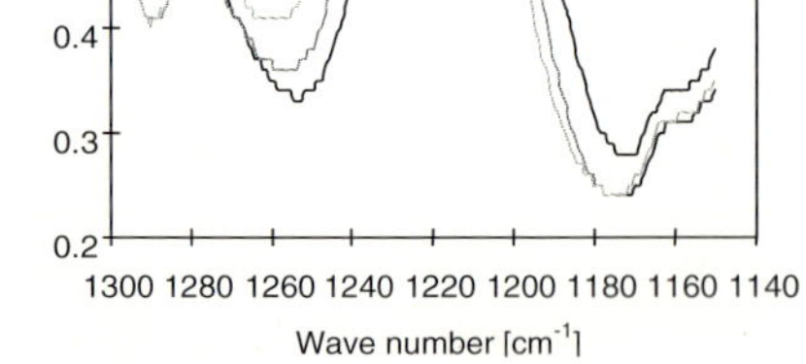

Fig. 2.6: FTIR-spectrum of the *Zn/D2EHPA* complex at different organic zinc concentrations

The phosphoryl bond is highly affected in the *Zn-D2EHPA* complex. The FTIR-spectrum of the zinc complex in n-heptane shows two concentration dependent $\nu\,(P{=}O)$ bands at 1234 cm^{-1} and 1206 cm^{-1}. In Fig. 2.6 can be seen, that the band of 1234 cm^{-1} decreases with increasing zinc concentration and the contrary can be seen at 1206 cm^{-1}. The first one represents thus the free

D2EHPA and the latter one the ion exchanger bound in the metal complex. The latter one is more difficult to interpret, since there are two ν (*P=O*) bands overlapping. This overlapping mechanism is given in Fig. 2.7. On the left side in Fig. 2.7 we see the bounding with the PO_2-anion and on the right side the bounding with an additional neutral *D2EHPA* molecule.

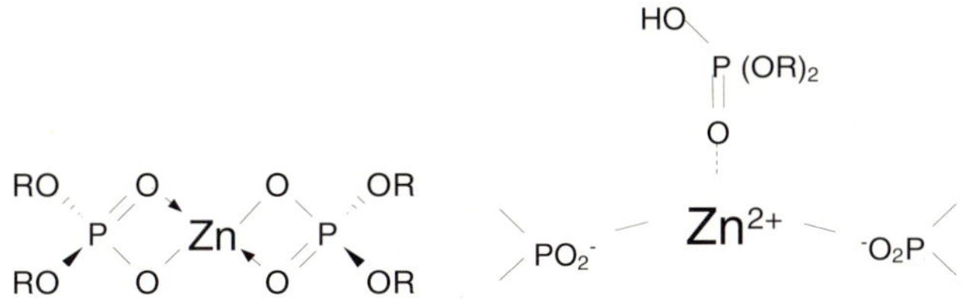

Fig. 2.7: Interaction between the *P–O* moieties with zinc

At a stoichiometric loading of 2:1 with no additional *D2EHPA* present, a complex according to the left structure in Fig. 2.7 is formed. The bond at 1234 cm^{-1} disappears and the only remaining ν (*P=O*) bond at 1198 cm^{-1} for a zinc complex (and 1197 cm^{-1} for a nickel complex) represents the *P=O* bond of the PO_2–anion (see Fig. 2.8).

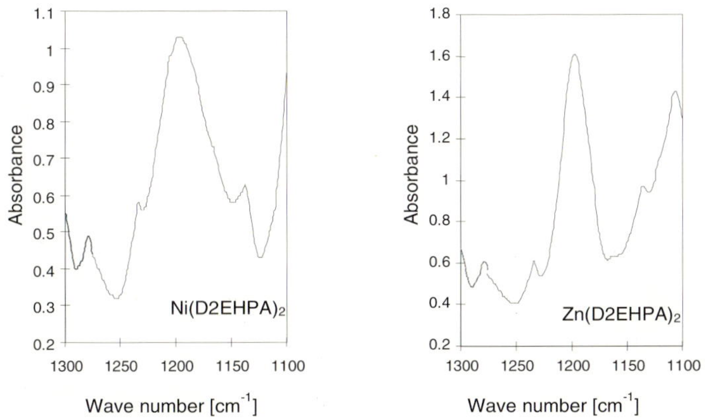

Fig. 2.8: ν (*P=O*) bound of a 2:1 *D2EHPA* / *Ni* respectively *Zn* complex.

However, at a lower zinc concentration this bound is at 1206 cm^{-1} , thus there is always an overlapping of the two mechanisms depicted in Fig. 2.7. For a quantification and analysis of the complex structure therefore the bond of the free *D2EHPA* at 1234 cm^{-1} should be used, as shown in Fig. 2.9. The result is a linear calibration curve in the range from 0.02 to 0.09 M *D2EHPA*. From a simple balance the concentration of the *D2EHPA* in the complex can be calculated:

$$[RH] = [RH]_0 - [RH]_{\text{free}} \tag{2.129}$$

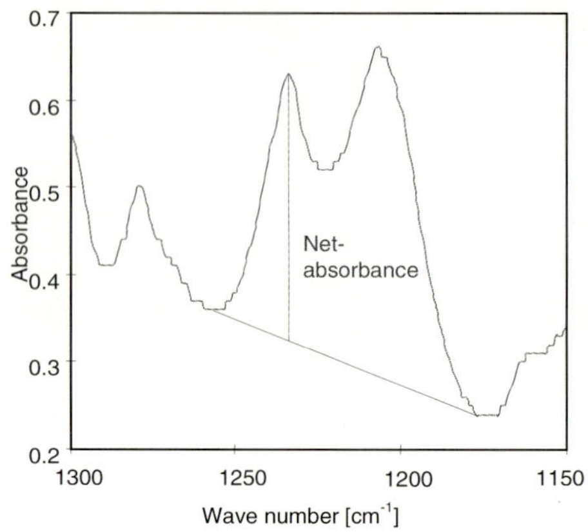

Fig. 2.9: Analysis of the free *D2EHPA* content at 1234 cm^{-1}

As can be seen in Fig. 2.10, the slope is three which gives the same result as the slope analysis according to equ. (2.128) respectively Fig. 2.3:

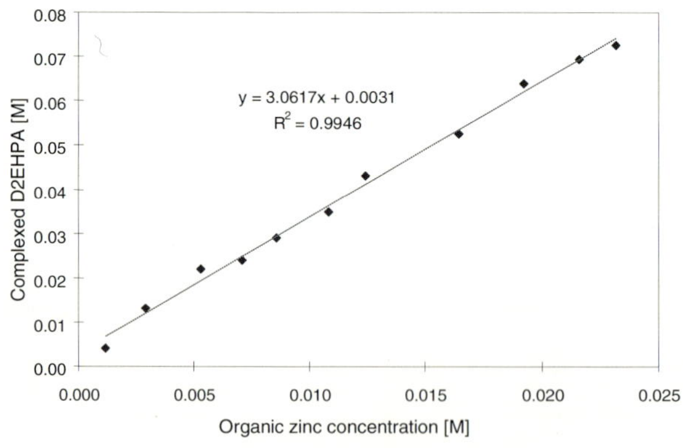

Fig. 2.10: FTIR-analysis of the complex stoichiometry

In a similar manner this can be performed with any other metal, e.g. sodium which reacts as:

$$Na^+ + c\overline{R_2H_2} \leftrightarrow \overline{NaR(RH)_{2 \cdot c-1}} + H^+ \quad K_{Na1,2c} \tag{2.130}$$

and the law of mass action in a logarithmic form is:

$$\log D_{Na} - \text{pH} = c \log\left(\overline{[R_2H_2]}\right) + \log K_{Na1,2c} \tag{2.131}$$

Since sodium is difficult to extract, the balance for *D2EHPA* reads in the whole concentration domain:

$$\overline{[R_2H_2]} = \overline{[R_2H_2]}_0 - c\,\overline{[NaR(RH)_{2\cdot c-1}]} \cong \overline{[R_2H_2]}_0 \tag{2.132}$$

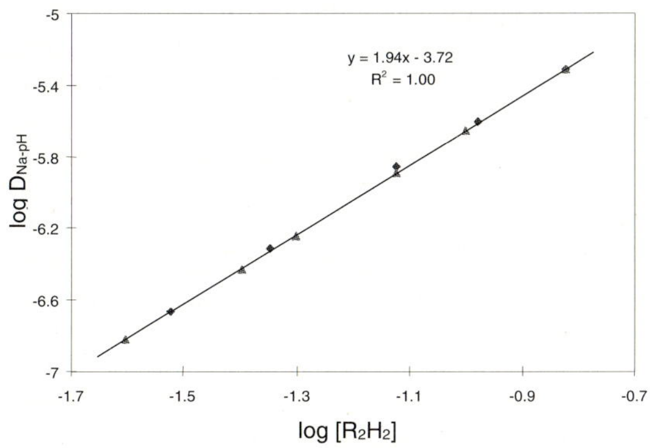

Fig. 2.11: Slope analysis of *Na-D2EHPA*

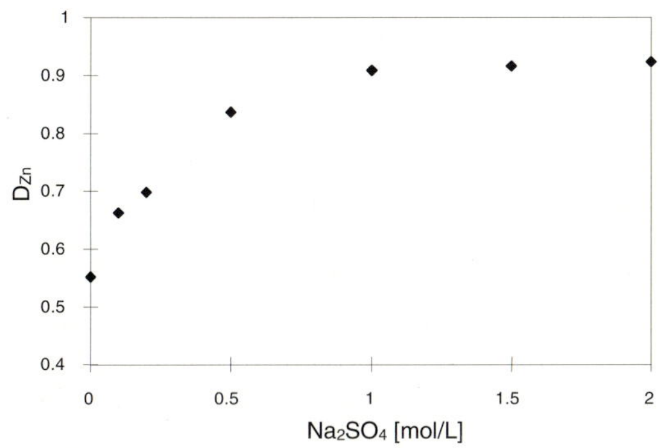

Fig. 2.12: Influence of sodium sulphate on the zinc distribution coefficient; $[ZnSO_4]_0 = 0.05$ [M], $[D2EHPA]_0 = 0.1$ [M]

The slope analysis in Fig. 2.11 reveals *c* to be 1.94, which gives a complex $NaR(RH)_3$:

$$Na^+ + 2\overline{R_2H_2} \leftrightarrow \overline{NaR(RH)_3} + H^+ \tag{2.133}$$

However, even if there is a negligible extraction of sodium the distribution coefficient of zinc is significantly affected in the presence of sodium sulphate, as depicted in Fig. 2.12.

The higher the aqueous sodium concentration is, the better is the zinc extraction, which may be due to the following phenomena:

- change of the aqueous activities,
- formation of mixed *Na/Zn/D2EHPA* complexes,
- H^+ puffer in the SO_4^{2-}/HSO_4^- system.

Which of these phenomena prevails is the result of the following theoretical analysis, based on the chemical potential theory in reactive extraction systems.

To calculate the activity coefficient γ, the Pitzer model [Pitzer & Mayorga (1973), Pitzer (1975, 1979), Pitzer et al. (1977)] is used for the aqueous phase and the Hildebrand-Scott (1950) solubility parameter is used for the organic phase. The Pitzer, Masson and solubility parameters are taken from the literature [Barton 1991]. With the software SXLSQI of Baes et al. (1990), the solubility parameter of the organic complex and the equilibrium constant can be estimated. The program is freely available via the Internet (http://www.ornl.gov/divisions/casd/csg/sxlsqi/). A list of all model parameters is given in the appendix E. The quality of the fit is expressed by the agreement factor σ (N_E is the number of the experimental data and N_P the number of adjustable parameters and σ_i denotes the experimental error):

$$\sigma = \left[\sum_{i=1}^{N_E} \left(\frac{1}{\sigma_i^2} (y_i - y_{i,\mathrm{calc}})^2 \right) \frac{1}{N_E - N_P} \right]^{0.5} \tag{2.134}$$

In the aqueous phase the following ions are present H^+, Zn^{2+}, HSO_4^-, SO_4^{2-} and also Na^+; the organic phase contains the complexes formed with *D2EHPA* in n-heptane. The reactions are summarised as follows. The extraction of zinc is:

$$Zn^{2+} + 1.5\overline{R_2H_2} \leftrightarrow \overline{ZnR_2RH} + 2H^+ \tag{2.135}$$

and the law of mass action on the basis of activities is then:

$$K_{1,3} = \frac{[\overline{ZnR_2RH}][H^+]^2}{[Zn^{2+}][\overline{R_2H_2}]^{1.5}} \frac{\gamma_{\overline{Zn}} \gamma_H^2}{\gamma_{Zn^{2+}} \gamma_{\overline{R_2H_2}}^{1.5}} \tag{2.136}$$

In the aqueous phase the hydrogen sulfate dissociation equilibrium reaction is valid:

$$H^+ + SO_4^{2-} \leftrightarrow HSO_4^- \tag{2.137}$$

with

$$K_{HS} = \frac{a_{HS}}{a_H \cdot a_S} = \frac{\gamma_{HS} \cdot [HSO_4^-]}{\gamma_H \cdot [H^+] \cdot \gamma_S \cdot [SO_4^{2-}]} = \frac{1}{0.0105} \tag{2.138}$$

If sodium is present then is:

$$Na^{+} + 2\overline{R_2H_2} \leftrightarrow \overline{NaR(RH)_3} + H^{+} \tag{2.139}$$

similarly:

$$K_{Na1,4} = \frac{[\overline{NaR(RH)_3}][H^{+}]}{[Na^{+}][\overline{R_2H_2}]^2} \frac{\gamma_{\overline{Na}}\,\gamma_H}{\gamma_{Na^+}\,\gamma^2_{\overline{R_2H_2}}} \tag{2.140}$$

For the zinc free Na_2SO_4 system the solubility parameter of the sodium complex was set equal to that of the solvent. Afterwards the equilibrium constant was estimated log $K_{N\,1,4} = -\,3.366 \pm 0.010$ with an agreement factor σ = 1.1531. For the sodium free $ZnSO_4$ system the solubility parameter of the dimeric *D2EHPA* was set equal to the solvent value; the equilibrium constant (log $K_{1,3} = -\,0.9441 \pm 0.088$) and the solubility parameter of the *Zn–D2EHPA* complex (δ = 18.591 ± 0.413 MPa$^{1/2}$) were estimated (σ = 1.6479). The agreement between the model simulations and experimental data can be seen in Fig. 2.13. Compared with the slope analysis presented in Fig. 2.3 only the data set in the diluted area (0.164 mM *Zn*) will follow the slope of 1.5, whereas in the higher concentrated region (0.15 M *Zn*) there is a deviation. Despite this deviation the model based on chemical potentials shows good resemblance. On the basis of these data, the performance of the mixed *Na/Zn*-system could be predicted as shown in Fig. 2.14 (σ = 0.96).

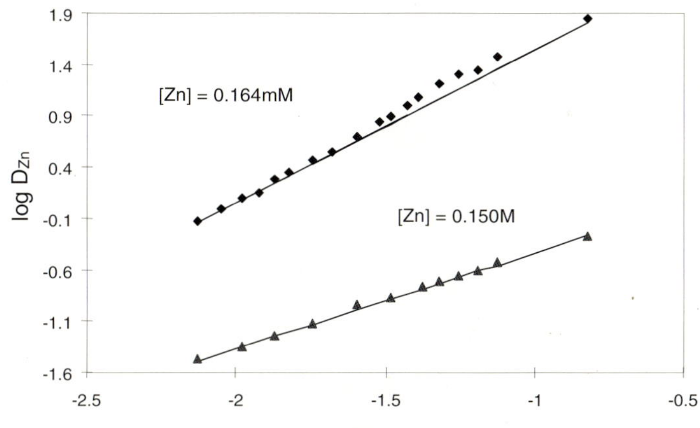

Fig. 2.13: Comparison of calculated and simulated equilibria of $ZnSO_4$

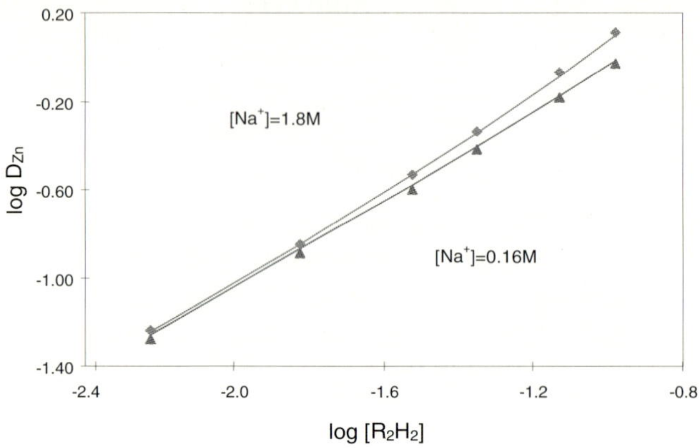

Fig. 2.14: Influence of sodium on the zinc extraction

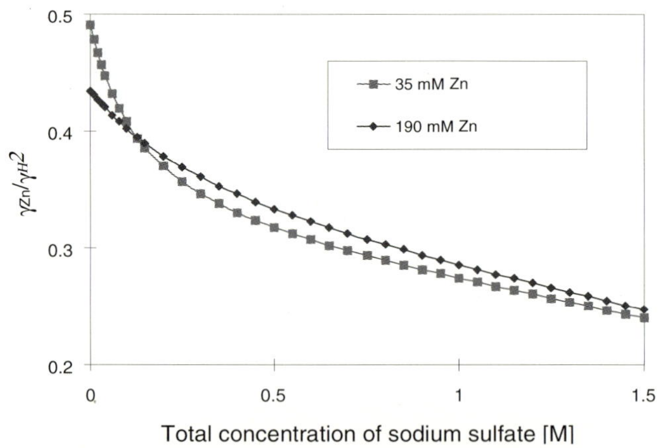

Fig. 2.15: γ_{Zn}/γ^2_H analysis with respect to equ. (2.136)

For that system the term γ_{Zn} / γ_H^2 versus the aqueous sodium sulfate (see Fig. 2.15) shows that the decrease of the activities will result in a reduced zinc extraction. However, since the contrary was expected and mixed *Na*/*Zn*-complexes could not be detected (see Fig. 2.14) the puffer capacity in the hydrogen sulfate system is the only reason for increases in the zinc distribution coefficient. The sulfate, which has its origin in Na_2SO_4, has the ability to compensate H^+ ions (see equ. (2.137)) and thus to raise the zinc distribution to a higher level (see Fig. 2.12). As can be seen, SO_4^{2-} and not Na^+ is the decisive species.

However, the advantage of using activity coefficients is to predict the system behaviour with other constituents from a binary data base. In a similar way as depicted here it could be shown that model parameters from the sulphate system allow a satisfactory prediction of the behaviour in the chloride system [Mörters & Bart (2000)], as is shown in Fig. 2.16 ($\sigma = 1.811$).

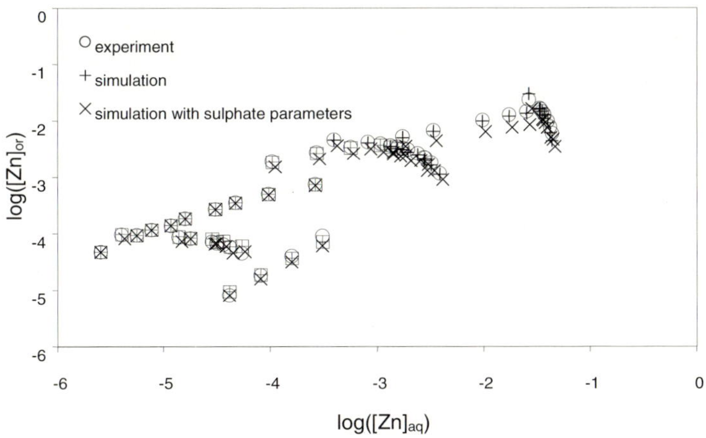

Fig. 2.16: Comparison of calculated and simulated equilibria in the chloride system

2.4.2 Acid/Anion Exchange System

The anion exchanger tri-n-octylamin (*TOA*) is also a well-known extractant for recovery of organic and inorganic acids and other anionic species. In a similar way to the previous section the extraction equilibria of acids in the presence of salts will be discussed. As an example, the extraction of acetic acid (*HAc*) and hydrochloric acid (*HCl*) with *TOA* in toluene in the presence of *NaCl* is considered. In the aqueous phase the following dissociation reactions will occur:

$$Na^+ + Cl^- \leftrightarrow NaCl \tag{2.141}$$

$$H^+ + Cl^- \leftrightarrow HCl \tag{2.142}$$

$$H^+ + Ac^- \leftrightarrow HAc \tag{2.143}$$

The dissociation constant, e.g. for acetic acid, is then:

$$K_{Diss} = \frac{[H^+][Ac^-]}{[HAc]} \frac{\gamma_{H^+}\gamma_{Ac^-}}{\gamma_{HAc}} \tag{2.144}$$

and reads similar for *HCl* and *NaCl*. The extraction reaction for e.g. acetic acid is then:

$$a\,HAc + b\,\overline{TOA} \leftrightarrow \overline{(TOA)_b(HAc)_a} \qquad (2.145)$$

with the equilibrium constant:

$$K_{ab} = \frac{[\overline{(TOA)_b(HAc)_a}]}{[(TOA)]^b[(HAc)]^a}\,\frac{\gamma_{\overline{(TOA)_b(HAc)_a}}}{\gamma_{TOA}^b\,\gamma_{HAc}^a} \qquad (2.146)$$

A slope analysis of the data by Kirsch et al. (1997) reveals at low concentrations (< 1 M *TOA*, < 0.2 M *HAc*) the formation of a 1:1 complex of *TOA* and *HAc*. At higher concentrations a 1:3 complex will develop according to the species distribution in Fig. 2.17.

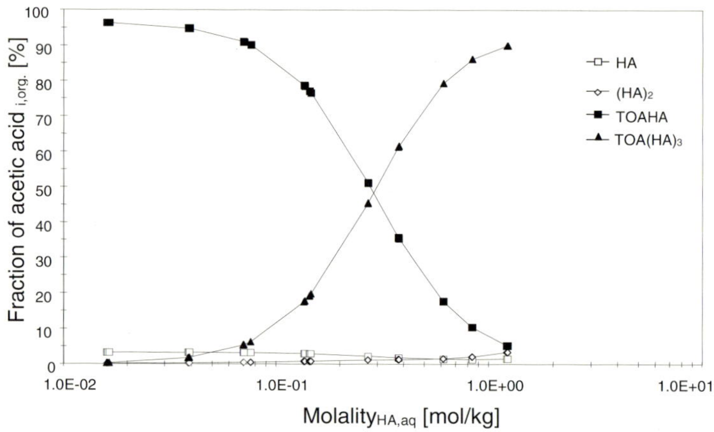

Fig. 2.17: Calculated fraction of organic acetic acid complexes for 10 wt.% *TOA* in the toluene system

As can be seen, the physically extracted *HAc* respectively $(HAc)_2$ dimer is on a very low concentration level and is negligible in most cases. As could be shown also with FTIR analysis, there is no water bound in the TOA–HAc complex, although there is an increasing solubility of water with higher acetic acid loadings (> 1 M) due to an increasing polarity of the organic phase [Ziegenfuß & Maurer (1994)]. All the thermodynamic parameters (Pitzer, Hildebrand–Scott solubility parameters etc.) are given in appendix E and Fig. 2.18 presents the phase equilibria for *HAc* in the presence of surfactants [Czapla 2000]. However, there is no influence of surfactants on the phase equilibria but this will change considering mass transfer. Fig. 2.19 depicts the influence of *NaCl* on the distribution of *HAc*. The prediction of the mixed system was only possible assuming the existence of a mixed complex $HCl(HAc)_2TOA$ [Roos & Bart (2000a)]. The model predicts accurately in the industrial relevant concentration domain, which is up to 1 M *TOA*. Higher ion exchanger concentrations are seldom used in industry due to viscosity reasons.

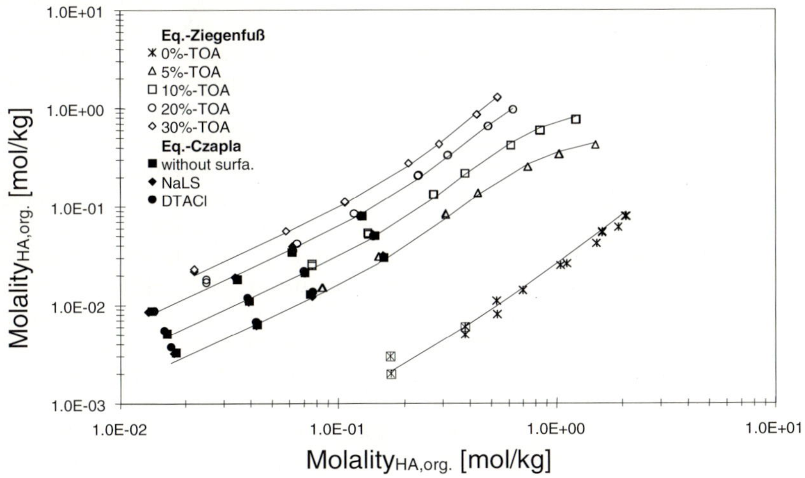

Fig. 2.18:. Comparison between measured and calculated acetic acid equilibrium data with TOA in toluene at 298K [Ziegenfuß & Maurer(1994), Czapla (2000)]

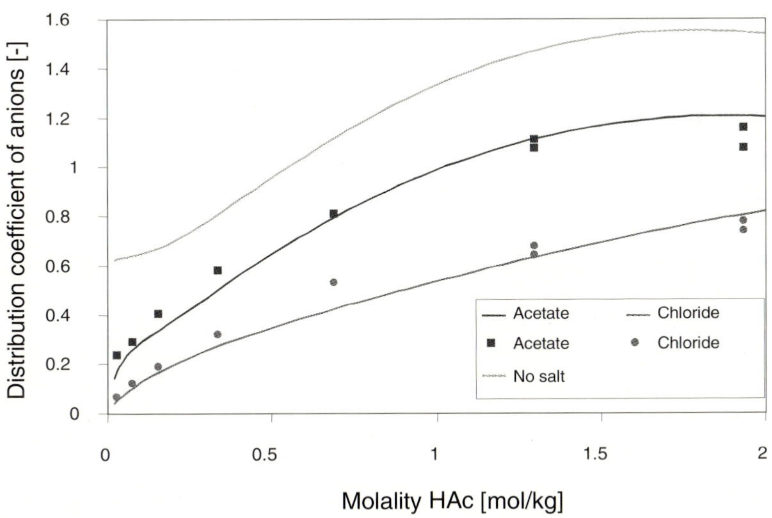

Fig. 2.19: Influence of *NaCl* on the extraction of *HAc* with *TOA* in toluene at 298K (0.55M NaCl, 20%TOA)

In a final summary we can conclude that the thermodynamic behaviour of ion exchange systems can be predicted from a binary data source. However, there is always the possibility that mixed complexes can occur in all systems especially with more aqueous components present and with increasing concentrations. The determination and evaluation of reactive phase equilibria data has always to be supplemented with slope analysis and spectroscopic data, in order to gain insight into the complexes respectively of the true species in each phase. The description on the basis of chemical potentials allows the prediction of the distribution behaviour, whether there is a change in diluent or inert salt (change from sulphate to chloride), etc. As has been shown here, the simple Hildebrand–Scott model, in which one parameter characterises a pure species, is very helpful in this respect. However, other thermodynamic approaches [Kirsch et al. (1997)] will lead to similar results and the quality is determined by the simplifications involved. Very often the role of water is considered to be negligible, which may often not be true.

The effect of surfactants, which may accumulate at a liquid-liquid interface, is of no significance for the bulk phase behaviour and thus for the thermodynamic equilibrium. However, ionic surfactants may induce a change separation and thus a electrostatic potential close to the interface, which may influence mass transfer of ionic species [Pfennig & Schwerin (1998), Czapla & Bart (2000)]. This will be discussed in detail in the next chapter. However, if the surfactant concentration exceeds the critical micelle concentration a situation will occur, which is usually not considered in classic liquid-liquid extraction. The use of such micellar extraction systems is discussed later in Chapter 4.

3 Reactive Mass Transfer

3.1 Introduction

In systems with liquid ion exchangers the following mass transfer regimes may exist:

- chemical reaction regime,
- diffusional controlled regime and
- mixed regime with both, diffusional and kinetic resistances.

The mass transfer with reactive extraction is vitally accomplished by the various chemical reactions occurring in the liquid ion exchanger and solute system. Usually, the solute (metal ion, inorganic acid, etc.) is insoluble in an organic environment and even when extracting organic soluble compounds (e.g. organic acids), the portion of physical extraction on the overall performance decreases till negligible. This statement is more valid when the solute concentration is less. However, liquid ion exchangers are industrially designed to dissolve preferably in the organic solvent phase (aqueous solubilities in the ppm-level) due to loss prevention, thus the reaction site is the organic to aqueous interface.

Most of the liquid ion exchangers have an amphiphilic structure. This enables a direct reaction of the polar, hydrophilic end with the solute. The lipophilic rest is in the organic environment and guarantees a high solubility and thus a high extraction capacity. These ligands are usually branched to enhance solubility and to reduce cristallization tendency. Additionally, the size of the ion exchanger should be moderate (molecular weight about 300 g/mol) otherwise the solubility decreases and sterical hindrances reduce interfacial kinetics considerably. However, the amphiphilic nature causes an adsorption of the ion exchanger species at the interface, which has to be taken into consideration in a proper kinetic model by using an adsorption isotherm.

The overall equilibrium reaction of a certain stoichiometry is approached in infinite time. This overall reaction can be divided into several elementary steps which may be either instantaneous or rate determining. These elementary steps may be

- the solute–ion exchanger complexation,
- the interfacial adsorption/desorption reaction,
- the aggregation or dissociation of the organic ion exchanger,
- the complexation reaction with chelating agents,
- the removal of solvating water molecules by an organic compound,

- the additional solvation of the 1:1 stoichiometry of the solute:ion exchanger complex to form higher complexes (e.g. 1:3),
- the competitive reaction of another solute,
- etc.

As a matter of consistency, the developed kinetic model that is based on these individual steps should approach the overall equilibrium reaction at infinite time, which is not always true when only initial rates are considered. From an engineering point of view the use of activities instead of concentrations in the kinetic equations is recommended in order to satisfy the final equilibrium conditions as has been discussed in Chapter 2.

The development of commercial ion exchangers has produced faster chemical kinetics performance, a higher capacity and lower losses (degradation, aqueous solubility, thermal stability, etc.). Added to this, the overall mass transfer hindrance may be by a rate-controlling diffusional process. The usually higher viscosity of the organic phase and the higher molecular weight of the produced organic complexes may lead to dominating diffusional resistances in the organic phase [Mörters & Bart (2000a)]. In the presence of surfactants this effect may be even enhanced. The enrichment of surfactants at the liquid-liquid interface is a daily experience in industrial practice. As a result the interface will be more rigid and difficult to be penetrated by the solute. These adsorption layers which lead to a restriction in diffusional mass transfer, have to be considered accordingly.

However, if the mass transfer is limited by an interfacial reaction, the restricted transport of the reactants to or of the products from the interface is no longer relevant. Then the adsorbate matrix and the induced charge at the interface have a marked effect on the interfacial kinetics of ionic solutes. Anionic surfactants will cause a negative charged interface and cationic surfactants a positive one. Neutral surfactants will behave indifferently. Anionic solutes will be repelled and cationic ones attracted by a negatively charged interface. These electrostatic interactions will thus change the solute concentrations at the interface which results in an enhanced or reduced chemical reaction rate. At high aqueous salt concentrations these effects should be less serious since the difference between the bulk and the interfacial charge concentrations diminishes and polarization effects are thus dampened. However, at high interfacial surfactant concentrations an interfacial blockage causing an even drastical reduction of the mass transfer rate may occur.

3.2 Physical Mass Transfer

3.2.1 Fick's Law

As we know from our working experience, the flux of a species i in a mixture with respect to that mixture is defined as

- a flux of a species i proportional to a concentration gradient, or
- a flux proportional to a concentration difference.

The first is known as Fick's law with the diffusivity, D, as a proportionality factor. The second law may be regarded as a rough difference form of the first, with a lumped parameter, k, the mass transfer coefficient, assuming diffusion takes place only in a thin film of length z adjacent to an interface:

$$k = D / z \tag{3.1}$$

Fick's law relates the diffusion flux J_I to the mole fraction gradient as to:

$$J_1 = -c_t \, D_{12} \, \nabla x_1 \tag{3.2}$$

The flux of component 2 can be obtained in the same way:

$$J_2 = -c_t \, D_{21} \, \nabla x_2 \tag{3.3}$$

Considering that both the diffusion fluxes and the mole fraction gradients sum up to zero it can be shown that the diffusion coefficients are symmetric in a binary case.

$$D_{12} = D_{21} = D \tag{3.4}$$

In a two component system there is only one independent diffusion flux driving force. Fick's diffusion coefficient is a function of temperature, pressure and composition but not of the driving force. In a multicomponent system there are n–1 independent mole fractions, since

$$\sum_{i=1}^{n} x_i = 1, \tag{3.5}$$

and n–1 independent driving forces, since

$$\sum_{i=1}^{n} \nabla x_i = 0, \tag{3.6}$$

and n–1 independent molar fluxes, since

$$\sum_{i=1}^{n} J_i = 0 \tag{3.7}$$

and

$$J_i = -c_t \sum_{k=1}^{n-1} D_{ik} \, \nabla x_k \tag{3.8}$$

Equation (3.8) is the generalized Fick's law of a multicomponent system which can be rewritten in the following n–1 dimensional matrix formulation:

$$(J) = -c_t \, [D] \, (\nabla x) \tag{3.9}$$

(J) represents a column matrix of order n–1. $[D]$ is a square matrix of dimension $(n-1) \times (n-1)$ and the column matrix (∇x) is of order n–1. The nondiagonal or off-diagonal elements or cross-coefficients D_{ik} ($i \neq k = 1,2,..., n-1$) are, in general, non-zero. They may be positive or negative and are in contrast to the binary case usually non-symmetric ($D_{12} \neq D_{21}$). Furthermore, the numerical values of D_{ik} depend on the particular choice of the system numbering and thus do not reflect the physical *i-k* interactions. Since the driving forces in the Fick's equations are concentration gradients, the non-idealities of the mixture are considered in the diffusion coefficients. As a result they depend strongly on the composition of the mixture. This situation is more relaxed in the case of the Maxwell–Stefan formulation as we will see later. Here the matrix of diffusivities is symmetric, the elements of which are always positive and independent of the chosen solvent component, and the non-idealities are considered in the driving force.

The molar flux, N_i, is determined by the velocity of a species (u_i) and is defined with respect to some outer reference frame (e.g. laboratory, interface, ...). However, the molar diffusion flux is defined with respect to an inner reference frame, e.g. the molar reference velocity of the mixture u:

$$J_i = c_i \, (u_i - u) \tag{3.10}$$

Other reference frames (mass, volume, ...) are possible [Cussler (1976)] and they can be transformed into each other. The mean velocity of a mixture contributes from the individual velocities to be:

$$u = \sum_{i=1}^{n} x_i \, u_i \tag{3.11}$$

This definition implies:

$$\sum_{i=1}^{n} J_i = 0 \tag{3.12}$$

However, we are not interested in fluxes with respect to a mixture but in fluxes with respect to a boundary, e.g. to an interface. In addition to the diffusion flux, there is also a drift flux towards this boundary which is defined as:

$$N_d = c_i \, u \tag{3.13}$$

The molar flux of a species, N_i, is then given by the sum of the diffusion flux and the drift flux:

$$N_i = J_i + c_i \, u = c_i \, u_i \tag{3.14}$$

Summation of the molar fluxes of all species delivers the total molar flux:

$$N_t = \sum_{i=1}^{n} N_i = c_t \, u \tag{3.15}$$

Combining equs. (3.14) and (3.15) yields the definition of the molar flux with respect to an outer reference frame:

$$N_i = J_i + \frac{c_i}{c_t} N_t = J_i + x_i \, N_t \tag{3.16}$$

In the one dimensional case we know this relation in the differential form as

$$N_i = -D_1 \frac{dc_i}{dz} + x_i N_t \tag{3.17}$$

or written in its difference form to be:

$$N_i = -k_i \, \Delta c_i + x_i \, N_t \tag{3.18}$$

Since we have n independent molar fluxes N_i but only n–1 independent diffusion fluxes J_i we need an additional equation, a so-called bootstrap relation, in order to solve the system of equations. An interesting case is the diffusion of a trace (or several traces) in a binary (or pseudo-binary) system whereas the bootstrap relation is given by the mole fraction of the pre-defined solvent component

$$x_2 \approx 1 \tag{3.19}$$

and hence

$$N_1 = -k_1 \Delta c_1 + (1 - x_2) \, N_t \tag{3.20}$$

As can be easily seen, the drift flux, also known as Stefan flux, tends to zero so that the molar flux equals the diffusion flux. From this special case we obtain our daily working equation, equ. (3.20). However, we will come back to the bootstrap conditions later on.

3.2.2 The Maxwell–Stefan Law

There are several fundamental theories from which a diffusion law can be derived theoretically. Fick (1855) developed his diffusion laws in analogy to Fourier's work [Fourier (1822)]. Stefan (1871) derived his equations after Maxwell (1866, 1952) taking a hydrodynamic approach into account when considering the friction between the species. This means that the molecular motion of a species i is counteracted by the friction forces of the species considered and the other constituents present in the mixture. For a binary mixture this yields in:

driving force of species 1 (d_1) = friction force between species 1 and 2 (F_R)

From a hydrodynamic point of view, the friction forces are proportional to the velocity differences:

$$F_R \propto (u_1 - u_2) \tag{3.21}$$

In order to consider the different masses of species, there is the necessity of introducing a weighting factor:

$$F_R = f_{12} \, x_1 \, x_2 \, (u_1 - u_2) \tag{3.22}$$

where f_{12} is the friction or drag coefficient. The inverse drag coefficient is defined as the Maxwell–Stefan diffusion coefficient $Đ_{12}$:

$$F_R = \frac{x_1\, x_2\, (u_1 - u_2)}{Đ_{12}} \tag{3.23}$$

The friction force on species 2 exerted by species 1 is defined analogously:

$$F_R = \frac{x_2\, x_1\, (u_2 - u_1)}{Đ_{21}} \tag{3.24}$$

According to Newton's principle the forces between 2 and 1 and 1 and 2 are identical which leads to symmetric binary Maxwell–Stefan diffusivities similar to Fick's case:

$$Đ_{12} = Đ_{21} \tag{3.25}$$

In a multicomponent mixture the friction forces exerted on species 1 sum up accordingly to the number of components, n, involved:

$$F_R = \sum_{j=1}^{n} \frac{x_1\, x_j\, (u_1 - u_j)}{Đ_{1j}} \tag{3.26}$$

For any species, i, this is:

$$F_R = \sum_{j=1}^{n} \frac{x_i\, x_j\, (u_i - u_j)}{Đ_{ij}} \tag{3.27}$$

The friction force equals the diffusional driving force, but with opposite sign:

$$d_i = -\sum_{j=1}^{n} \frac{x_i\, x_j\, (u_i - u_j)}{Đ_{ij}} \tag{3.28}$$

The generalised driving force, d_i, can be derived from the theory of irreversible thermodynamics [Taylor, Krishna (1993)] and is generally defined as:

$$c_t\, RT\, d_i = c_i\, \nabla_{T,P}\mu_i + (\varphi_i - \omega_i)\nabla P - \rho_i \left(\overline{F}_i - \sum_{j=1}^{n} \omega_j\, \overline{F}_j \right) \tag{3.29}$$

The first term on the right side represents a chemical potential gradient at constant pressure and temperature. In contrast to the generalised Fick's law the chemical potential gradient replaces the concentration gradient. If there is a difference in volume fraction φ_i and weight fraction w_i of a species i, the pressure gradient (second term) may play a role (e.g. centrifuge). The last term becomes important in electrical fields, e.g. with electrolytes, which will be discussed in detail in the following subchapter.

For all external forces the following is valid:

$$\overline{F}_i = \frac{F_i}{M_i} \tag{3.30}$$

where F_i is the external body force per mol of species i. In absence of external forces the chemical potential is the only driving force for non-electrolytes:

$$d_i = \frac{c_i}{c_t} \frac{1}{RT} \nabla_{T,P} \mu_i = \frac{x_i}{RT} \nabla_{T,P} \mu_i \tag{3.31}$$

For an ideal fluid, where the activity coefficient equals one and with

$$\mu_i = \mu_i^0 + RT \ln(x_i) \tag{3.32}$$

the driving force term reduces to:

$$d_i = \nabla x_i \tag{3.33}$$

For this ideal system the Maxwell–Stefan equation reads:

$$\nabla x_i = -\sum_{j=1}^{n} \frac{x_i \, x_j \, (u_i - u_j)}{Đ_{ij}} \tag{3.34}$$

Together with equ. (3.14) this can be rewritten into molar fluxes

$$\nabla x_i = \sum_{j=1}^{n} \frac{(x_i \, N_j - x_j \, N_i)}{c_t \, Đ_{ij}} \tag{3.35}$$

or in terms of molar diffusion fluxes to give:

$$\nabla x_i = \sum_{j=1}^{n} \frac{(x_i \, J_j - x_j \, J_i)}{c_t \, Đ_{ij}} \tag{3.36}$$

The generalised Maxwell–Stefan law for a multicomponent system considering any driving force is then:

$$d_i = \sum_{j=1}^{n} \frac{(x_i \, N_j - x_j \, N_i)}{c_t \, Đ_{ij}} \tag{3.37}$$

or

$$d_i = \sum_{j=1}^{n} \frac{(x_i \, J_j - x_j \, J_i)}{c_t \, Đ_{ij}} \tag{3.38}$$

However, note that the Maxwell–Stefan diffusion coefficient $Đ_{ij}$ is by definition a friction coefficient between the species i and j. That implies that the D_{ii} ($Đ_{11}$, $Đ_{22}$, ...) are thus not defined. The exact formulation is then:

$$d_i = \sum_{\substack{j=1 \\ j \neq i}}^{n} \frac{(x_i \, J_j - x_j \, J_i)}{c_t \, Đ_{ij}} \tag{3.39}$$

but for $i = j$ the numerator equals zero, solving that problem.

Due to the summation condition for diffusion fluxes (equ. (3.12)) only n–1 diffusion flux equations are independent analogous to the generalised Fick's law. The flux of the *n*th component can be calculated by rearranging equ. (3.12).

$$J_n = -\sum_{j=1}^{n-1} J_j \tag{3.40}$$

Combing the equs. (3.38) and (3.40) yields:

$$\begin{aligned} c_t\, d_i &= \left\{ \sum_{\substack{j=1 \\ j\neq i}}^{n-1} \frac{\left(x_i\, J_j - x_j\, J_i\right)}{Đ_{ij}} \right\} - \frac{x_n\, J_i}{Đ_{in}} - \frac{x_i \sum_{j=1}^{n-1} J_j}{Đ_{in}} \\ &= \sum_{\substack{j=1 \\ j\neq i}}^{n-1} \frac{x_i\, J_j}{Đ_{ij}} - J_i \sum_{\substack{j=1 \\ j\neq i}}^{n-1} \frac{x_j}{Đ_{ij}} - J_i \frac{x_n}{Đ_{in}} - \sum_{\substack{j=1 \\ j\neq i}}^{n-1} \frac{x_i\, J_j}{Đ_{in}} - \frac{x_i\, J_i}{Đ_{in}} \\ &= -J_i \sum_{\substack{j=1 \\ j\neq i}}^{n} \frac{x_j}{Đ_{ij}} - J_i \frac{x_i}{Đ_{in}} + \sum_{\substack{j=1 \\ j\neq i}}^{n-1} x_i \left(\frac{1}{Đ_{ij}} - \frac{1}{Đ_{in}} \right) J_j \end{aligned} \tag{3.41}$$

This can be written in a more compact form by introducing the coefficients B_{ii} and B_{ij}:

$$c_t\, d_i = -B_{ii}\, J_i - \sum_{\substack{j=1 \\ j\neq i}}^{n-1} B_{ij}\, J_j \tag{3.42}$$

where B_{ii} and B_{ij} are defined as:

$$B_{ii} = \frac{x_i}{Đ_{in}} + \sum_{\substack{k=1 \\ k\neq i}}^{n-1} \frac{x_k}{Đ_{ik}} \tag{3.43}$$

$$B_{ij} = -x_i \left(\frac{1}{Đ_{ij}} - \frac{1}{Đ_{in}} \right) \tag{3.44}$$

As an example, for a ternary system this results in:

$$B_{1,1} = \frac{x_1}{Đ_{1,3}} + \frac{x_2}{Đ_{1,2}} + \frac{x_3}{Đ_{1,3}}\ ,\ B_{1,2} = -x_1 \left(\frac{1}{Đ_{1,2}} - \frac{1}{Đ_{1,3}} \right) \tag{3.45}$$

and

$$B_{2,1} = -x_2 \left(\frac{1}{Đ_{2,1}} - \frac{1}{Đ_{2,3}} \right),\ B_{2,2} = \frac{x_2}{Đ_{2,3}} + \frac{x_1}{Đ_{2,1}} + \frac{x_3}{Đ_{2,3}} \tag{3.46}$$

A mathematically more elegant form of equ. (3.42) can be obtained by matrix formulation:

$$c_t\,(d) = -[B]\,(J) \tag{3.47}$$

To express this in (J) the inverse of $[B]$ is needed:

$$c_t [B]^{-1}\,(d) = -[B]^{-1}\,[B]\,(J) \tag{3.48}$$

With $[B]^{-1} \cdot [B]$ results the unity matrix $[I]$ and thus:

$$(J) = -c_t\,[B]^{-1}\,(d) \tag{3.49}$$

For an ideal mixture d_i can be replaced by ∇x_i considering concentration gradients as the only driving force present in the system:

$$(J) = -c_t\,[B]^{-1}\,(\nabla x) \tag{3.50}$$

A comparison with the generalised Fick's law (equ. (3.9)):

$$(J) = -c_t\,[D]\,(\nabla x) \tag{3.9}$$

gives the relation between the diffusivities:

$$[B]^{-1} \equiv [D] \tag{3.51}$$

As can be seen in ideal binary systems the two diffusivities are identical:

$$\frac{1}{B_{11}} = Đ_{12} = D_{12} = D \tag{3.52}$$

For non-ideal fluids activity coefficient models have to be considered in the driving force calculation:

$$\mu_i = \mu_i^0 + RT\,\ln(x_i \gamma_i) \tag{3.53}$$

The total differential is at constant pressure and temperature a function of the composition:

$$\mathrm{d}\mu_i = \sum_{j=1}^{n} \left(\frac{\partial \mu_i}{\partial x_j} \right)_{P,T,x_k \neq x_j} \mathrm{d}x_j \tag{3.54}$$

and leads to

$$d_i = \frac{x_i}{RT} \sum_{j=1}^{n-1} \left(\frac{\partial \mu_i}{\partial x_j} \right)_{P,T,x_k \neq x_j} \nabla x_j \tag{3.55}$$

or in mole fraction gradients:

$$
\begin{aligned}
d_i &= \frac{x_i}{RT}\sum_{j=1}^{n-1} RT\left(\frac{\partial \ln(x_i\,\gamma_i)}{\partial x_j}\right)_{P,T,x_k\neq x_j} \nabla x_j \\
&= x_i \sum_{j=1}^{n-1}\left\{\left(\frac{\partial \ln x_i}{\partial x_j}\right)+\left(\frac{\partial \ln \gamma_i}{\partial x_j}\right)\right\}_{P,T,x_k\neq x_j} \nabla x_j \\
&= \sum_{j=1}^{n-1}\left\{\delta_{ij} + x_i\left(\frac{\partial \ln \gamma_i}{\partial x_j}\right)_{P,T,x_k\neq x_j}\right\}\nabla x_j \\
&= \sum_{j=1}^{n-1}\Gamma_{ij}\,\nabla x_j
\end{aligned}
\qquad (3.56)
$$

whereas δ_{ij} is the Kronecker delta and Γ_{ij} the thermodynamic factor defined by:

$$
\Gamma_{ij} = \delta_{ij} + x_i\left(\frac{\partial \ln \gamma_i}{\partial x_j}\right)_{P,T,x_k\neq x_j} \qquad (3.57)
$$

For an ideal fluid is $\Gamma_{ii} = 1$ and $\Gamma_{ij} = 0$.

For non-ideal mixtures equ. (3.49) is extended by the matrix of thermodynamic factors:

$$
(J) = -c_t\,[B]^{-1}\,[\Gamma]\,(\nabla x) \qquad (3.58)
$$

The relation between the two diffusion coefficients in a non-ideal binary mixture is now:

$$
\frac{\Gamma_{11}}{B_{11}} = Đ_{12}\,\Gamma_{11} = D_{12} \qquad (3.59)
$$

In a binary system only one independent Maxwell-Stefan or Fick diffusion coefficient exists:

$$
Đ_{12}\,\Gamma = D_{12} = D \qquad (3.60)
$$

On can solve this equation by using the Fick diffusion coefficient to calculate the Maxwell–Stefan diffusivity. The case of a multicomponent system is given below:

$$
[B]^{-1}\,[\Gamma] = [D] \qquad (3.61)
$$

In this way it is possible to calculate Maxwell–Stefan diffusivities from binary Fick diffusivities in conjunction with appropriate g^E models and vice versa.

3.2.3 Electrolyte Systems

In electrolyte systems the driving force has to be extended to account also for the electrostatic forces occurring in the system.

$$c_t\ RT\ d_i = c_i\ \nabla_{T,P}\mu_i + (\varphi_i - \omega_i)\nabla P - c_i\left(F_i - \omega_i \sum_{j=1}^{n} c_j\ F_j\right) \tag{3.62}$$

These are defined as [Newman (1991)]:

$$F_i = -z_i\ \Im\ \nabla\phi \tag{3.63}$$

$\Im$ is the Faraday constant (9.65×10^4 C/mol) and ϕ denotes the electrical potential. Not considering pressure gradients equ. (3.62) can be rewritten as:

$$c_t\ RT\ d_i = c_i\ \nabla_{T,P}\mu_i + \left(c_i\ z_i - \omega_i \sum_{j=1}^{n} c_j\ z_j\right)\Im\ \nabla\phi \tag{3.64}$$

and together with the electroneutrality condition

$$\sum_{i=1}^{n} c_i\ z_i = 0 \tag{3.65}$$

this simplifies to

$$c_t\ RT\ d_i = c_i\ \nabla_{T,P}\mu_i + c_i\ z_i\ \Im\ \nabla\phi \tag{3.66}$$

In analogy to non-electrolytes we use the molar reference velocity which is that of the solvent u_n:

$$J_i^n = N_i - c_i\ u_n \tag{3.67}$$

The superscript n denotes the chosen reference frame. With that choice of reference frame the diffusion flux of the solvent is

$$J_n^n = 0 \text{ and } \sum_{i=1}^{n} J_i^n \neq 0 \tag{3.68}$$

The Maxwell–Stefan formulation is now given by:

$$d_i = \sum_{j=1}^{n} \frac{\left(x_i\ J_j^n - x_j\ J_i^n\right)}{c_t\ Đ_{ij}} \tag{3.69}$$

Introducing again the B_{ij} into the Maxwell–Stefan formulation of the chosen reference frame

$$c_t\ d_i = -\sum_{j=1}^{n-1} B_{ij}^n\ J_j^n \tag{3.70}$$

with

$$B_{ii}^{n} = \sum_{\substack{k=1 \\ k \neq i}}^{n-1} \frac{x_k}{Ð_{ik}} \quad i = 1,2,\ldots,n-1 \tag{3.71}$$

$$B_{ij}^{n} = -\frac{x_i}{Ð_{ij}} \quad i = 1,2,\ldots,n-1 \tag{3.72}$$

allows to rewrite this equation in matrix form. Equ. (3.73) gives the solution solved for the diffusion fluxes

$$(J^{n}) = -c_t \cdot [B^{n}]^{-1} \cdot (d) \tag{3.73}$$

However, ionic–ionic diffusion cross coefficients D_{ij} are scarcely investigated. In diluted systems the ion–water interaction always dominates. For dilute solutions one yields the Nernst–Planck equation when using only the diagonal elements of the matrix $[B_n]$ with

$$B_{ii}^{n} = \frac{1}{Ð_{in}^{0}} = \frac{1}{D_i} \qquad B_{ij}^{n} = 0 \quad (i \neq j) \tag{3.74}$$

The Maxwell–Stefan diffusion coefficient of an ionic species i in water $Ð_{in}^{0}$ is furtheron written as D_i which is the diffusion coefficient at infinite dilution. The driving force of a diluted, ideal system can be obtained from equ. (3.66).

$$d_i = \nabla x_i + x_i \, z_i \, \frac{\Im}{RT} \nabla \phi \tag{3.75}$$

This expression can also be converted into concentration units:

$$c_t \, d_i = \nabla c_i + c_i \, z_i \, \frac{\Im}{RT} \nabla \phi \tag{3.76}$$

In this system the diffusion flux is then:

$$J_i^{n} = -D_i \nabla c_i - c_i \, z_i \, D_i \frac{\Im}{RT} \nabla \phi \tag{3.77}$$

Together with equ. (3.67) one obtains the most common form of the Nernst–Planck equation:

$$N_i = -D_i \nabla c_i - c_i \, z_i \, D_i \frac{\Im}{RT} \nabla \phi + c_i \, u_n \tag{3.78}$$

Here one can see that the three terms on the right-hand side, namely diffusion, migration and convection, contribute to the molar flux N_i. The additional electroneutrality constraint can be differentiated to give:

$$\sum_{i=1}^{n} z_i \nabla c_i = 0 \tag{3.79}$$

This means that the total system is described by n–2 independent composition gradients. The sum in equ. (3.79) includes all ionic species and the solvent. However, the solvent is without charge which gives a trivial solution.

For further convenience a new variable n_I is introduced which denotes the number of ionic species. With only one solvent this number is $n_I = n-1$. The motion of ionic species induces a component-specific current:

$$z_i \, N_i \, \Im \tag{3.80}$$

The sum of these single currents gives the current of the mixture:

$$i = \Im \sum_{j=1}^{n_I} z_j \, N_j \tag{3.81}$$

whereas the N_j are given by the Nernst–Planck equation:

$$i = -\Im \sum_{j=1}^{n_I} z_j \, D_j \, \nabla c_j - \frac{\Im^2}{RT} \nabla \phi \sum_{j=1}^{n_I} c_j \, z_j^2 \, D_j \tag{3.82}$$

The last term of the Nernst–Planck equation ($u_n \cdot c_i$) can be omitted due to the electroneutrality condition. The following definitions are common:

- equivalent conductivity of species j:

$$\kappa_j = \frac{\Im^2}{RT} c_j \, z_j^2 \, D_j \tag{3.83}$$

- equivalent conductivity of the mixture:

$$\kappa = \sum_{j=1}^{n_I} \kappa_j = \frac{\Im^2}{RT} \sum_{j=1}^{n_I} c_j \, z_j^2 \, D_j \tag{3.84}$$

- the transference number of species j:

$$t_j = \frac{\kappa_j}{\kappa} \tag{3.85}$$

This definition lets one convert equ. (3.82) into:

$$i = -\Im \sum_{j=1}^{n_I} z_j \, D_j \, \nabla c_j - \kappa \, \nabla \phi \tag{3.86}$$

and can be rearranged to:

$$\nabla \phi = -\frac{i}{\kappa} - \frac{\Im}{\kappa} \sum_{j=1}^{n_I} z_j \, D_j \, \nabla c_j \tag{3.87}$$

As can be seen, even when there is no current applied (i = 0) a finite electrical potential exists which is termed diffusion potential. With this it is possible to substitute the potential in the migration term in the Nernst–Planck equation:

$$N_i = -D_i \nabla c_i + c_i u_n + c_i z_i D_i \frac{\Im}{RT}\frac{i}{\kappa} + c_i z_i D_i \frac{\Im^2}{RT}\frac{1}{\kappa}\sum_{j=1}^{n_I} z_j D_j \nabla c_j \tag{3.88}$$

or in a simpler form using the transference numbers:

$$N_i = -D_i \nabla c_i + c_i u_n + \frac{t_i}{z_i \Im} i + \frac{t_i}{z_i}\sum_{j=1}^{n_I} z_j D_j \nabla c_j \tag{3.89}$$

The nth gradient in the above summation is replaced by the differentiated electroneutrality equ. (3.79):

$$N_i = -D_i \nabla c_i + c_i u_n + \frac{t_i}{z_i \Im} i + \frac{t_i}{z_i}\sum_{j=1}^{n_I-1} z_j \left(D_j - D_n\right)\nabla c_j \tag{3.90}$$

In order to apply the Nernst–Planck equation also for moderately diluted mixtures, one has to use activity coefficients:

$$N_i = -D_i \nabla c_i - D_i c_i \nabla \ln \gamma_i^n - c_i z_i D_i \frac{\Im}{RT}\nabla \phi + c_i u_n \tag{3.91}$$

The "neutral" activity coefficient γ_i^n can be derived from single ion activities as follows:

$$\ln \gamma_i^n = \ln \gamma_i - \frac{z_i}{z_n}\ln \gamma_n \tag{3.92}$$

However, the choice of the reference ionic species is arbitrary [Newman (1991)]. For a non-ideal mixture the Nernst–Planck equation yields:

$$N_i = -D_i \cdot \nabla c_i - D_i \cdot c_i \cdot \nabla \ln \gamma_i + c_i \cdot u_n + \frac{t_i}{z_i \cdot \Im}\cdot i + \frac{t_i}{z_i}\sum_{j=1}^{n_I-1} z_j \left(D_j - D_n\right)\nabla c_j + \frac{t_i}{z_i}\sum_{j=1}^{n_I} z_j c_j D_j \nabla \ln \gamma_j \tag{3.93}$$

In case of "pure" diffusional processes like ion exchange or solvent extraction, there is no current in the system:

$$i = 0 \tag{3.94}$$

When extracting ionic species by means of liquid ion exchangers without co-extracting water, then the solvent velocity is zero

$$u_n = 0 \tag{3.95}$$

and hence equ. (3.93) simplifies to:

$$N_i = -D_i \nabla c_i - D_i c_i \nabla \ln \gamma_i + \frac{t_i}{z_i} \sum_{j=1}^{n_l - 1} z_j \left(D_j - D_n\right) \nabla c_j + \frac{t_i}{z_i} \sum_{j=1}^{n_l} z_j c_j D_j \nabla \ln \gamma_j \tag{3.96}$$

3.2.4 Bootstrap Relation

In one of the preceding sections we have derived the (n–1) independent diffusion flux equations of a multicomponent system. Since we have n components and therefore n independent molar fluxes, one needs a bootstrap condition to solve the entire system. In liquid–liquid reactive extraction systems there are two cases of practical importance, the stagnant species and the reaction stoichiometry.

Stagnant species. If a bulk species is not extracted then its velocity, u_n, is zero. This case is known as Stefan diffusion, when a trace species performs a fast diffusion through a stagnant film [Cussler (1997)]. This is exactly true for an ideal solvent extraction system where the organic diluent is virtually immiscible with water and only the solute/-s will pass the interface. In many textbooks this is considered as a special diffusion phenomenon, whereas with the Maxwell–Stefan formulation only the bootstrap condition must be correctly set.
For a species, where u_n equals zero, there is:

$$N_n = J_n + x_n N_t = 0 \tag{3.97}$$

Rearranging this equation leads to an expression for the total molar flux:

$$N_t = \frac{-J_n}{x_n} \tag{3.98}$$

The fluxes of the remaining species, i, are

$$N_i = J_i + x_i N_t = J_i - \frac{x_i J_n}{x_n} = \left(1 + \frac{x_1}{x_n}\right) J_i + \frac{x_i}{x_n} \sum_{\substack{k=1 \\ k \neq i}}^{n-1} J_k \tag{3.99}$$

Reactions. If the fluxes are constrained by a chemical reaction with a certain stoichiometry (note: educts have a negative sign) the bootstrap relation is given by:

$$N_i \frac{1}{\nu_i} = N_j \frac{1}{\nu_j} \tag{3.100}$$

The total flux is then:

$$N_t = \sum_{j=1}^{n} N_j = \sum_{j=1}^{n} N_i \frac{\nu_j}{\nu_i} = \frac{N_i}{\nu_i} \sum_{j=1}^{n} \nu_j \tag{3.101}$$

Together with the general definition of the molar flux, N_i can be written as:

$$N_i = J_i + x_i\, N_t = J_i + x_i \frac{N_i}{\nu_i} \sum_{j=1}^{n} \nu_j \tag{3.102}$$

or solved for N_i:

$$N_i = \frac{J_i}{1 - \frac{x_i}{\nu_i} \sum_{j=1}^{n} \nu_j} \tag{3.103}$$

In reactive liquid–liquid extraction a part of the species will diffuse in the aqueous or in the organic phase. Thus for species in phase α this reads as:

$$(N_i)_\alpha = \frac{(J_i)_\alpha}{1 - \frac{x_i}{\nu_i} \left(\sum_{j=1}^{n} \nu_j \right)_\alpha} \tag{3.104}$$

and respectively in phase β:

$$(N_k)_\beta = \frac{(J_i)_\beta}{1 - \frac{x_i}{\nu_k} \left(\sum_{j=1}^{n} \nu_j \right)_\beta} \tag{3.105}$$

However, there are other bootstraps possible (for example equimolar exchange or a stagnant membrane) but not likely in reactive extraction. As further readings to this topic the book of Taylor and Krishna (1993) is recommended.

3.2.5 Estimation of Diffusion Coefficients

As has been mentioned above, in case of an ideal binary mixture there is no difference between the Fick or the Maxwell–Stefan diffusion coefficient. For non-ideal binary mixtures a relation between the two has already been derived

$$Đ_{12} \Gamma_{11} = D_{12} \tag{3.59}$$

Since the thermodynamic non-idealities of the mixture are concentrated in the thermodynamic correction factor Γ the Maxwell–Stefan diffusion coefficients are less concentration dependent than those of Fick as has also been shown experimentally. This is essentially true for gases and with more restrictions also for liquids [Wesselingh, Krishna (2000)]. Liquid systems at infinite dilution behave as ideal mixtures. Diffusion coefficients at infinite dilution D^0

are then equal to the Maxwell–Stefan ones. If species 1 is highly diluted, then the diffusivity is:

$$Đ_{12}^{1\to 0} = Đ_{21}^{1\to 0} = D_{12}^{0} \tag{3.106}$$

or on the contrary:

$$Đ_{12}^{2\to 0} = Đ_{21}^{2\to 0} = D_{21}^{0} \tag{3.107}$$

In the multicomponent case the relation between the Fick and the Maxwell–Stefan diffusion coefficient has been shown earlier as:

$$[B^{-1}][\Gamma]=[D] \tag{3.61}$$

The non-idealities in the Maxwell–Stefan formulation are considered in the thermodynamic factor. However, in concentrated multicomponent mixtures the liquid diffusion coefficients depend strongly on composition (and even on viscosity). A summary of the binary mixing rules that are necessary for the prediction of diffusivities between the two extremes at infinite dilution is given by Pertler et al. (1996). The interpolation scheme proposed by Vignes (1966) is recommended by Reid et al. (1988). For a binary it reads:

$$Đ_{12} = Đ_{21} = (D_{12}^{0})^{x_2}\ (D_{21}^{0})^{x_1} \tag{3.108}$$

It was for a ternary system first suggested by Cullinan and Cusik (1967) and reads for n components:

$$Đ_{ij} = \left(D_{ij}^{0}\right)^{x_j} \left(D_{ji}^{0}\right)^{x_i} \prod_{\substack{k=1 \\ k\neq i,j}}^{n} \left(D_{ij,x_k\to 1}\right)^{x_k} \tag{3.109}$$

In a ternary system one needs nine parameters in order to determine the three Maxwell–Stefan coefficients ($Đ_{12}$, $Đ_{13}$, $Đ_{23}$). The six binary ones can be easily obtained, since they are equal to the binary diffusivities at infinite dilution (D_{ij}^{0} D_{ji}^{0}). The three ternary diffusivities of the type $D_{ij}^{k\to one}$ are difficult to obtain. These represent the interaction of i and j in almost pure k. As an example how one can estimate this value, Wesselingh and Krishna (2000) suggest the use of a geometric average of the binary diffusivities for the "i–j" pair at infinite dilution.

$$D_{ij,x_k\to 1} = \left(D_{ij}^{0}\ D_{ji}^{0}\right)^{\frac{1}{2}} \tag{3.110}$$

Fortunately, the influence of this parameter on ternary flux calculations is not so strong. As x_i and x_j are almost zero, the friction of dilute "i" and dilute "j" in almost pure "k" is negligible compared with the total friction [Rutten (1992)]. However, there are other mixing rules [Taylor & Kooijman (1991)], but the weighting factors or exponents must sum to unity. In any case, for a ternary flux calculation (see Fig. 3.1) the following boundaries of equ. (3.109) must be valid:

$$\lim_{x_j\to 1} Đ_{ij} = D_{ij}^{0} \tag{3.111}$$

$$\lim_{x_i \to 1} Đ_{ij} = D^0_{ji} \tag{3.112}$$

and

$$\lim_{x_i \to 0} Đ_{ij,x_j=0} = \lim_{x_j \to 0} Đ_{ij,x_i=0} \tag{3.113}$$

The last condition is indicated by the arrows in Fig. 3.1. Following the binary border line, one must end at $x_3 = 1$. However, the experimental data basis on ternary diffusion coefficients is scarce [Rutten (1992), Thiel (1995), von Reden (1998), van de Ven-Lucassen (1999)] and due to technical problems regarding the experimental determination of diffusivities at these boundaries one is constrained to the existing mixing rules and predictive equations. A comparison of experimental techniques to determine diffusion coefficients is given elsewhere [Tyrell & Harris (1984), Cussler (1976)] and will not be discussed here.

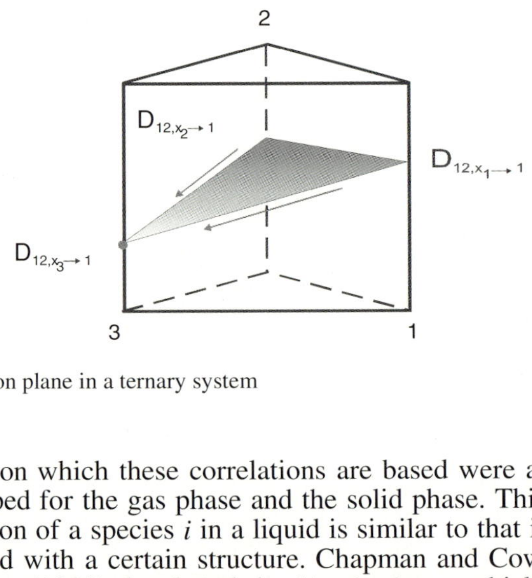

Fig. 3.1: D_{12} diffusion plane in a ternary system

The theories on which these correlations are based were adapted from the theories developed for the gas phase and the solid phase. This means that the diffusional motion of a species *i* in a liquid is similar to that in a dense gas or in a molten solid with a certain structure. Chapman and Cowling (1952) and Enskog and Sven (1922) developed the *kinetic theory* which is valid for the prediction of diffusivities in gases:

$$D = \frac{1.86\times10^{-3} T^{3/2} (1/\overline{M}_1 + 1/\overline{M}_2)^{1/2}}{p\sigma_{12}^2 \Omega} \tag{3.114}$$

where D is the diffusion coefficient in cm²/s, T is the absolute temperature in Kelvin, p is the pressure in atmospheres and M_i are the molecular weights. σ_{12} and Ω are molecular properties which are tabulated elsewhere [Hirschfelder et al. (1954)].

For liquids, an empirical modification of Enskog's equation was proposed by Chen et al. (1983):

$$D_{12} = h_{12} (D_{12}^{SHS} / D_{12}^E) D_{12}^E \tag{3.115}$$

where h_{12} is the rotational factor accounting for angular momentum transfer upon collision, (D_{12}^{SHS} / D_{12}^{E}) is the ratio of a smooth hard-sphere to the Enskog diffusivity. The factor accounts for dynamically correlated motion such as backscattering and vortex formation and D_{12}^{E} is the Enskog diffusivity in the dense gas region.

In the *activated jump theory* a movement of molecules occurs when after a certain 'activation energy' an intermediate 'transition state' of diffusion is found. According to a statistical distribution of the thermal energy only a small part of the molecules in the lattice can move. This assumption makes it possible to calculate an "equilibrium constant" for a pseudo-equilibrium between normal and activated molecules by elementary statistical thermodynamic methods. When this calculation is coupled with the quasi-lattice model of liquids, the diffusivity can be described as the product of two terms, one temperature independent and one that shows an exponential dependency on the temperature. According to Eyring's theory [Eyring & Hirschfelder (1937)], the diffusivity in liquids at infinite dilution can be calculated as follows:

$$D_{AB}^{0} = \frac{\lambda^2}{V_f^{1/3}} \left(\frac{kT}{2\pi m}\right)^{1/2} exp\left(-\frac{E}{RT}\right) \tag{3.116}$$

where λ is the elementary "jump distance" within the range of the intermolecular distance, V_f is the molecular volume, m the mass of the molecule, k the Bolzmann constant, T the temperature in Kelvin, R the gas constant, and E is the diffusion activation energy.

The *free volume theory* is based on the fact that only 55–80% of the liquid volume is occupied by the molecules. The volume between the molecules of a liquid is called the "free volume" or "holes". These voids are of different size and shape due to the random motion of the molecules. A molecule can only move if the void adjacent to it is large enough. The temperature dependence is similar for all liquids. Guggenheim (1935) suggested:

$$\frac{V^*}{V_c} = [1 + 1.75(1 - \frac{T}{T_c})^{1/3} + 0.75(1 - \frac{T}{T_c})]^{-1} \tag{3.117}$$

where V_c and T_c are the critical volume and temperature. At T=0 K this equation yields a minimum (or compressed) volume:

$$V^* = 0.268\, V_c \tag{3.118}$$

At higher temperatures and lower pressures the fluid expands and the "free volume" is:

$$V_F^* = V - V^* \tag{3.119}$$

where V is the molar volume. Only the void fraction of the holes larger than the molecules considered is of interest since through these the diffusion process can occur. Cohen and Turnbull (1959) developed a statistical model and estimated this fraction as:

$$\varphi = \exp[-0.7\frac{V^*}{V_F^*}] \tag{3.120}$$

The Stokes–Einstein theory is based on a hydrodynamic approach [Einstein (1905)]:

$$D = \frac{k_B T}{f} = \frac{k_B T}{6\pi\eta R_0} \tag{3.121}$$

where f is the frictional coefficient of the solute, k_B is the Boltzmann constant, η is the solvent viscosity, R_0 is the solute radius and the T is temperature in K. This equation is accurate to only about 20 % [Reid et al. (1988)], but remains as a standard to judge other correlations. When this equation is used for large species surrounded by small solvent molecules, the result seems reasonable. But when the solute size is less than five times of the solvent, this approach fails because this correlation is derived assuming a rigid solute sphere diffusing in a continuum (Stokes law). Unfortunately, the molecules of many commercial solvents are usually larger than those of the solutes. Therefore, several modifications of the Stokes–Einstein equation have been proposed. Rutten (1992) has developed an empirically modified equation to consider the diffusivity of a mixture consisting of two species with different diameters.

$$D_{21} = \frac{k_B T}{n_{SE}^0 \pi\eta R_2^2 / R_1} \tag{3.122}$$

where R_2 and R_1 are the radii of the solute and solvent respectively, n^0_{SE} is a constant in the range from 3.53 to 4.94 depending on the system when R_2/R_1 is varied between 0.7 to 1.6.

The *nonequilibrium molecular dynamics method* (NEMD) for computing transport coefficients (shear viscosity, thermal conductivity and diffusivity) of fluids based on the knowledge of binary interaction between molecules were discussed by Cummings and Evans (1992) and Sunderrajan et al. (1996). The motion of several thousands of molecules was simulated computationally, and from these simulations the diffusivities could be evaluated. There is some success for gases and it may develop to an ultimate method to calculate liquid diffusivities. However, NEMD methods have been seldom applied to compute transport properties in mixtures of industrial importance. Even the estimation of self-diffusivities in simple systems is erroneous. For example, the comparison of experimental data and estimated values from NEMD for CO_2 at 313.0 K results in a relative error up to 83%. The method might have a future with increasing computer power. Nowadays it is only possible to handle about 50000 atoms in reasonable time but 10^6 would be necessary to describe a continuum, but progress with gas–solid systems is reported [Keil (1999)].

However, all these theoretically derived correlations to predict diffusivities in liquids are strongly idealised which leads to doubtful results. In order to overcome this lack of accuracy *semi-empirical correlations* were developed [Reid et al. (1988)]. The most popular one is the correlation proposed by Wilke and Chang (1955):

$$D_{AB} = \frac{7.4 \times 10^{-8} (\varphi M_B)^{0.5} T}{\eta V_A^{0.6}} \tag{3.123}$$

where D_{AB} = diffusivity of A diluted in solvent B, cm^2/s, M_B = molecular weight of the solvent, T = temperature, K, μ = viscosity of the solution, cP, V_A = molar volume of the solute at normal boiling point, cm^3/g mol, ϕ = association factor of the solvent (ϕ =2.6 for water as solvent, 1.9 for methanol, 1.5 for ethanol, 1.0 for non-associated solvents such as benzene or ethyl ether).

Depending on the type of solvent–solute system Hayduk and Minhas (1982) proposed several rather similar correlations which have been obtained by regression analysis. For non-aqueous solutions it reads as:

$$D_{AB} = 1.55 \times 10^{-8} \frac{T^{1.29} (p_B^{0.5} / p_A^{0.42})}{\eta_B^{0.92} V_B^{0.23}} \tag{3.124}$$

with

$$p = V\,\sigma^{1/4} \tag{3.125}$$

Equ. (3.124) is similar to that of Tyn and Calus (1975) which is:

$$D_{AB} = 8.93 \times 10^{-8} \left(\frac{V_A}{V_B^2} \right)^{1/6} \left(\frac{p_B}{p_A} \right)^{0.6} \frac{T}{\eta_B} \tag{3.126}$$

Another equation has been proposed by Nakanishi (1978):

$$D_{AB} = \left[\frac{9.97 \times 10^{-8}}{(I_A V_A)^{1/3}} + \frac{2.4 \times 10^{-8} A_B S_B V_B}{I_A S_A V_A} \right] \frac{T}{\eta_B} \tag{3.127}$$

where I_A and S_A are solute and A_B and S_B are solvent specific parameters.

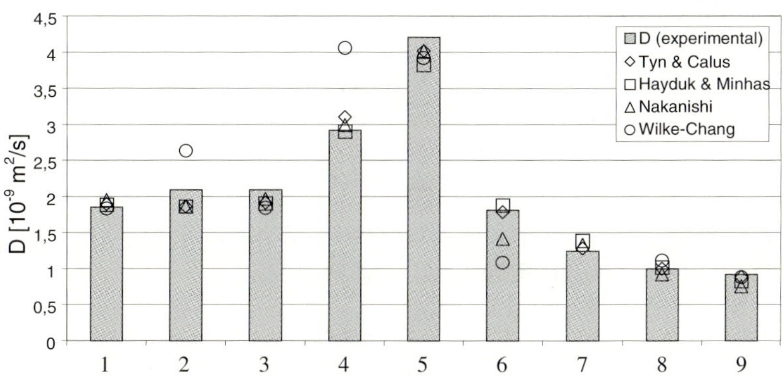

Fig. 3.2: Comparison of estimated and experimental values (1 toluene in benzene, 2 acetic acid in benzene, 3 cyclohexane in benzene, 4 acetic acid in acetone, 5 toluene in n-hexane, 6 benzene in ethanol, 7 water in ethanol, 8 ethanol in water, 9 aniline in water)

However, all correlations can be applied restrictively due to temperature, viscosity and structure of the compounds. Therefore it is not possible to extrapolate from the defined domain. Within the given physical boundaries the correlations behave quite well, as can be seen from Fig. 3.2.

In order to achieve higher flexibility, Quayle (1953) tabulated structural group contributions to calculate P in equ. (3.124). However, the group contribution concept is highly promising and was already applied successfully for the prediction of equilibrium thermodynamics with the UNIFAC model [Fredenslund et al. (1975)]. On the basis of this idea Fei and his co-workers [Ye (1988), Guo et al. (1997), Ye et al. (1996)] proposed a new practical approach to predict diffusion coefficients.

The *functional group contribution* method is based on the following assumptions:

- The diffusion process is governed by the diffusion areas S_A, S_B of the molecules A and B and their activation energies E.
- Both the diffusion areas and the activation energies are calculated from the contributions of each group which compose the molecules.

The correlations were derived from the Eyring equation and the model parameters from experimental data. In this way they obtained for the diffusivity at infinite dilution:

$$D^0_{AB} = k_0 T^{1/2} \exp(-\frac{E}{RT}) \tag{3.128}$$

where k_0 is defined as

$$k_0 = K\ S_A^a\ S_B^b \tag{3.129}$$

As can be seen from the equation the original parameters λ and V_f which are difficult to determine are substituted by k_0 which is solely a function of the diffusion areas. The molecule diffusion area is calculated from the summation of the functional group diffusion areas.

$$S_A = \sum_{k=1}^{m} \nu_{AK} Q_k; \qquad S_B = \sum_{k=1}^{n} \nu_{BK} Q_k; \tag{3.130}$$

where Q_k is the group diffusion area, ν_{AK}, ν_{BK} are the numbers of a functional group Q_k in molecule A and B, respectively. With

$$\frac{D^0_{AB}}{D^0_{BA}} = \left(\frac{S_A}{S_B}\right)^{a-b} \tag{3.131}$$

the following relation can be obtained:

$$a - b = 1.10 \tag{3.132}$$

when correlating experimental diffusivity data of hydrocarbons. The self-diffusion coefficient D_s for hydrocarbons is then:

$$D_s = K\, S^{a+b}\, T^{1/2} \exp(-\frac{E}{RT}) \tag{3.133}$$

Based on a large amount of self-diffusivities found in the literature the parameter K and the summation of a and b were estimated to be:

$$a + b = 0.8 \tag{3.134}$$

$$K = 2.0 \times 10^{-5} \tag{3.135}$$

Combining the equs. (3.132) and (3.134) one obtains a = 0.95 and b = –0.15. Thus the correlation for diffusivity at infinite dilution is:

$$D_{AB}^{\infty} = \phi 2.0 \times 10^{-5} \frac{S_A^{0.95}}{S_B^{0.15}} T^{0.5} \exp(-\frac{E}{RT}) \tag{3.136}$$

(association factor ϕ = 8 for H_2O, ϕ = 2 for alcohols except methanol, otherwise $\phi = 1$).

For ideal solutions the diffusion area of pseudo-solute and solvent are calculated as:

$$S_1 = x_B S_A + x_A S_B \,;\, S_2 = x_A S_A + x_B S_B \tag{3.137}$$

Substitution of S_1 and S_2 in equ. (3.136) yields the diffusivity of an ideal solution:

$$D_{id} = \phi\, K \frac{(x_B S_A + x_A S_B)^{0.95}}{(x_A S_A + x_B S_B)^{0.15}} T^{1/2}\, exp(-E/RT) \tag{3.138}$$

For non-ideal solutions, the thermodynamic correction factor is introduced:

$$\Gamma = 1 + \frac{d \ln \gamma_A}{d \ln x_A} \tag{3.139}$$

thus

$$D = \Gamma D_{id} \tag{3.140}$$

Here γ_A is the activity coefficient which is calculated with any g^E-model or the UNIFAC model. The diffusion activation energies are defined as

$$E_{AB} = \sum_{i=1}^{n} \sum_{j=1}^{n} \nu_{Ai} \nu_{Bj} E_{AiBj} \tag{3.141}$$

where E_{AiBj} is the diffusional activation energy between the group i in component A and group j in component B. Since $E_{AiBj} = E_{BiAj}$, it follows $E_{AB} = E_{BA}$.

By regressing a large amount of experimental data of self-diffusivities and binary diffusion coefficients at infinite dilution, the molecular diffusion area and the group diffusion activation energies could be estimated from equs. (3.137) and (3.141). Some examples of molecular diffusion areas are listed in

Table 3.1; the molecular diffusion activation energies of these groups are summarised in Table 3.2.

Table 3.1: Some examples of molecular group diffusion areas

Group	Q	Examples	$S=v_j Q_j$
CH_3	0.848	n-heptane	4.396
CH_2	0.540	n-dodecane	7.096
$[CH_2]^*$	1.292	cyclo-hexane	7.75
ACH^{**}	1.179	benzene	7.06
$C_6H_5CH_3$	5.168	toluene	5.168
$C_6H_5CH_2$	4.457	n-butyl-benzene	6.385

Table 3.2: Some examples of group diffusion activation energies

CH_3	851.4					
CH_2	312.00	-27.67				
$[CH_2]$	461.79	101.06	333.58			
ACH	667.40	58.00	306.58	294.44		
$C_6H_5CH_3$	3395.1	423.18	1799.7	1746.8	10038.0	
$C_6H_5CH_2$	3395.1	423.18	1676.75	1746.8	10038.0	10038.0
Group	CH_3	CH_2	$[CH_2]$	ACH	$C_6H_5CH_3$	$C_6H_5CH_2$

* [] .....cyclo, ** A .... aromatic

More than 600 experimental data from the literature and further experimental data of about 150 systems in wide temperature ranges were used to verify the model. The deviations of the group contribution method are less than 5% for most systems and less than 10% for few aqueous and alcohol systems. The comparison between this method and some other methods are presented in the articles by Fei and Bart (1998, 2000).

3.2.6 Mass Transfer

The mass transfer concept is derived considering the flux of a species with respect to a concentration difference. The corresponding mass transfer coefficient for a binary system is defined as follows [Bird et al. (1960)]:

$$k_B = \lim_{\substack{N_1 \to 0 \\ N_2 \to 0}} \frac{N_{1,B} - x_{1,B} N_t}{c_t \left(x_{1,B} - x_{1,I} \right)} = \lim_{\substack{N_1 \to 0 \\ N_2 \to 0}} \frac{J_{1,B}}{c_t \, \Delta x_1} \left[\frac{m}{s} \right] \tag{3.142}$$

Higher fluxes need a finite flux correction Ξ, since the flow of the mixture distorts the velocity and concentration profiles near the interface. The zero flux mass transfer coefficient k_B and the high flux mass transfer coefficient $k_B^\bullet$ are related in the following way:

$$k_B^\bullet = \Xi\, k_B \tag{3.143}$$

Definitions of the high flux correction factor are given elsewhere [Bird et al. (1960), Taylor & Krishna (1993)]. However, the concentration dependency of the mass transfer coefficients may be neglected in most cases for reactive extraction which was confirmed by Klocker (1996). Analogous to equ. (3.9) we find (n–1) fluxes with a symmetric mass transfer coefficient matrix with $(n-1) \times n/2$ coefficients:

$$(J_B) = (N) - (x_B)\, N_t = c_t\, [k_B]\,(x_B - x_1) = c_t\, [k_B]\,(\Delta x) \tag{3.144}$$

Steward and Prober (1964) and Toor (1964) proposed independently a method to simplify the mathematics involved. In the so-called linearised theory it is assumed that both the total molar concentration c_t and the matrix of diffusion coefficients $[Đ]$ are constant along the diffusion path. Furthermore $[Đ]$ is regarded as concentration independent. A further simplification made by Krishna (1977) was that also the matrix of thermodynamic factors $[\Gamma]$ is constant and is calculated at the average composition between bulk, B, and interface, I:

$$[\Gamma_{av}] = [\Gamma] \text{ at } (x_{av}) = \frac{(x_I + x_B)}{2} \tag{3.145}$$

respectively:

$$[D_{av}] = [B_{av}]^{-1}\,[\Gamma_{av}] = [B]^{-1}\,[\Gamma] \text{ at } (x_{av}) = \frac{(x_I + x_B)}{2} \tag{3.146}$$

As has been shown [Steward & Prober (1964), Arnold & Toor (1967), Taylor & Webb (1980)] these assumptions are justified since the deviations from the exact solution are negligible. A further simplification was introduced by Wesselingh and Krishna (2000), when approximating the driving force in a difference form. As we know the exact formulation is:

$$d_i = \frac{c_i}{c_t}\frac{1}{RT}\nabla_{T,P}\mu_i = \frac{x_i}{RT}\nabla_{T,P}\mu_i \tag{3.31}$$

and in combination with equ. (3.32) one yields for an ideal solution

$$\Delta\mu_i = \mu_i - \mu_i^0 = RT \ln x_i \tag{3.147}$$

For given T and p in one dimension is:

$$\frac{\mathrm{d}\mu_i}{\mathrm{d}z} = \frac{RT}{x_i}\frac{\mathrm{d}x_i}{\mathrm{d}z} \tag{3.148}$$

and in a linearised approximation according to Wesselingh and Krishna (2000):

$$\frac{\Delta \mu_i}{\Delta z} = \frac{RT}{x_{i,\mathrm{av}}} \frac{\Delta x_i}{\Delta z} \tag{3.149}$$

The Maxwell–Stefan equation for the ideal system reads (compare equ. (3.34)):

$$\frac{\Delta x_i}{x_{i,\mathrm{av}}} = \sum_{j=1}^{n} \frac{x_j (u_j - u_i)}{(Đ_{ij} / \Delta z)} \tag{3.150}$$

or in terms of molar fluxes (compare equ. (3.35)):

$$x_{i,av} \frac{\Delta x_i}{x_{i,av}} = \sum_{j=1}^{n} \frac{x_i N_j - x_j N_i}{c_t \, (Đ_{ij} / \Delta z)} \tag{3.151}$$

Probably due to error compensation, this approximation formula for the effect of composition change has been often found to give better results than the exact formula. However, all these simplifications help to easier solve the numerics of the system of the (n–1) independent flux equations.

As was shown initially, the mass transfer coefficient is a lumped parameter and according to the film model of Lewis & Whitman (1924) it is defined as:

$$k = D / z \tag{3.152}$$

For the penetration theory of Higbie (1935) there is:

$$k = \sqrt{\frac{4\,D}{\pi\, t_e}} \tag{3.153}$$

where t_e is the exposure time of a molecule at the interface. Alternatively, there is the surface renewal theory of Danckwerts (1951):

$$k = \sqrt{D\, s} \tag{3.154}$$

with the surface renewal frequency s. However, for technical applications there are correlations available derived from a dimensional analysis or founded on a empirical basis. A review on this is given by Slater (1994). In extractor design one major problem arises concerning an accurate description of mass transfer in droplet systems. The behaviour of droplet swarms (coalescence and wake effects, etc.) as well as the mobility of the drops are not fully understood and therefore difficult to predict [Wesselingh & Bollen (1999)]. For small drops (usually below 2 mm) we have a rigid spherical drop. When the drop size increases internal circulation becomes important, as the droplet rising velocity increases. At a certain diameter (very system dependend, ≥ 6 mm) the droplets start to oscillate until they break up into smaller ones. All this is naturally affected by the interfacial tension and the surfactants adsorbed on the interface. The hydrodynamics is influenced when lowering the surface tension and enhancing the emulsification tendency. Also the difference in the slip velocity of surfactant free and laden droplets [Frumkin & Levich (1947)] is a well known fact.

The influence of surfactants on the mass transfer can be seen in different aspects. The basic physical process by which a surface tension gradient induces convection in the bulk phase is known [Marangoni (1871)] as the "Marangoni effect" [Thomson (Lord Kelvin) (1871)]. Here, the surface tension variations caused by local inhomogenities in either the surface temperature or the solute concentration create the instability which perceptibly enhances mass transfer. In a theoretical analysis of Sterling and Scriven (1959) on the basis of the original Boussinesq theory (1905) it was demonstrated that interfacial viscosity effects (surfactant layers) will dampen the interfacial instability. Further reading on this topic can be found in the articles by Ortiz (1992) and Edwards et al. (1991).

Continuous phase. The mass transfer in the continuous phase can be regarded as a steady-state process due to the assumption of an infinite surrounding. This process can be divided into two extreme cases, the mass transfer to rigid drops and to drops with fully developed internal motion. The first one describes the lower boundary to mass transfer and is characterised by the creeping flow condition. A theoretical derivation of this results in:

$$Sh_c = 1 + (1 + Pe)^{1/3} \tag{3.155}$$

where $Pe = Re \cdot Sc$. For $Pe \to 0$ one reaches the limit for molecular diffusive transport ($Sh = 2$). The upper boundary was first derived by Boussinesq (1913) and represents the potential flow condition:

$$Sh_{c,crit} = \frac{2}{\sqrt{\pi}} \mathrm{Pe}_c^{0.5} \tag{3.156}$$

In this case the continuous phase velocity equals the surface velocity. Besides these theoretical derivations several empirical correlations were developed based on the following general form:

$$Sh_c = 2 + c_1 \, Re^{c_2} \, Sc^{c_3} \tag{3.157}$$

Steiner (1986) introduced the following empirical correlation in which solely the continuous phase mass transfer resistance is considered. For a rigid drop it is:

$$Sh_{c,R} = 2.43 + 0.773 \, Re_d^{0.5} \, Sc_c^{0.33} + 0.0103 \, Re_d Sc_c^{0.33} \tag{3.158}$$

For circulating drops the following critical limit is:

$$Sh_{c,crit} < Sh_{c,R} \tag{3.159}$$

and

$$Sh_{c,crit.} = \frac{2}{\sqrt{\pi}} Pe_c^{0.5} \tag{3.156}$$

respectively:

$$\frac{Sh_c - Sh_{c,R}}{Sh_{c,crit} - Sh_{c,R}} = 1 - \exp\ (-0.00418\ Pe_c^{0.42}) \tag{3.160}$$

Dispersed phase. In the dispersed phase the mass transfer is of a more complex nature than in the continuous phase since a steady-state process can no longer be assumed. Additionally, the different flow patterns inside the drop – stagnant, laminar circulating and turbulent circulating – as well as the possibility of drop oscillation have to be taken into account. Dependent on the internal motion of the drop the following theoretical mass transfer correlations have been published in the literature. For very small liquid drops with $Re_d < 1$ the mass transfer in the drop occurs primarily by molecular diffusion. Assuming a constant continuous phase concentration and infinite continuous phase mass transfer coefficients, an approximation of the diffusion equation gives [Newman (1931a,b)]:

$$\mathrm{Sh}_d = \frac{k_d d_p}{D} = 6.58\ \frac{D}{d_p} + 0.33 \frac{d_p}{4\, t_e} \tag{3.161}$$

As the drop size increases ($\mathrm{Re}_d < 10$) circulation within the drop takes place and mass transfer rates increase due to internal circulation and increased velocities. Kronig and Brink's (1950) solution gives

$$Sh_d = 17.9 \frac{D}{d_p} + 0.071 \frac{d_p}{t_e} \tag{3.162}$$

At higher contact times a mass transfer coefficient results which is 2.72 times higher than the limiting case of the rigid drops. Note that the mass transfer coefficient decreases as the drop diameter increases. Combined with a lower surface area per unit volume, the rate of mass transfer decreases substantially.

When the drops become larger, the circulation in the drops becomes turbulent. Handlos and Baron (1957) approximated the mass transfer coefficient to be

$$k_d = 0.00375\ w_d \left(\frac{\eta_c}{\eta_c + \eta_d} \right) \tag{3.163}$$

Note that the mass transfer coefficient becomes independent of the diffusion coefficient (and eliminates diffusion interactions inside the drop)! Despite the fact that its derivation is based on an incorrect assumption (cylindrical instead of spherical flow patterns), the Handlos–Baron equation and variations of it are widely used to predict dispersed phase mass transfer coefficients.

The equations presented above are those which are commonly used. There are also modifications of these, e.g. Steiner (1986) used the Newman equation with an empirical eddy diffusivity, E, which modifies the molecular diffusivity D as to:

$$\frac{E}{D} = 1 + 0.177\,\mathrm{Re}^{0.43} \left[\frac{\eta_c}{\eta_c + \eta_d} \right]^{0.89} Sc_d^{0.23} \tag{3.164}$$

This equation is valid for *E/D* < 10 and, for droplet swarms, a similar correlation is available. All these equations are only valid in clean systems. Since contaminants are always present in technical systems Slater (1995) introduced the stagnant cap model, which considers contamination, with respect to surfactants. Especially, liquid ion exchangers used in reactive extraction are of amphiphilic nature and will adsorb at the interface (see Fig. 3.3).

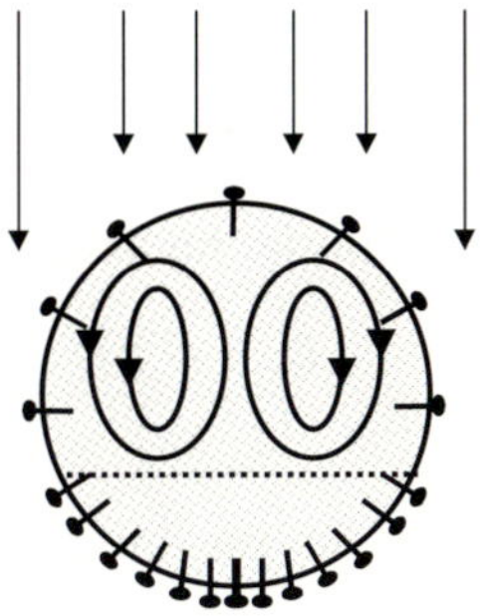

Fig. 3.3: Stagnant cap model

In the stagnant cap model the droplet is divided in a circulating part and one that is stabilised by the adsorbed surfactants. Based on Lochiel's (1965) improvement of the equation for potential flow by introducing the term $\alpha^{0.5}$ with

$$\alpha = 1 - \frac{2 + 3\,(1 + \eta_d / \eta_c) + m}{1 + (\rho_d * \eta_d / \rho_c\, \eta_c)^{0.5}} \, \frac{1.45}{Re^{0.5}} \tag{3.165}$$

which accounts for contamination. Slater has extended the range of validity ($10 < Re < 1000$) of this equation by introducing the approximation $\alpha = 1 - x \approx (1 + x)^{-1}$. In this way α is modified to be

$$\alpha = \left[1 + \frac{2 + 3\,(1 + \eta_d / \eta_c) + m}{1 + (\rho_d * \eta_d / \rho_c\, \eta_c)^{0.5}} \, \frac{1.45}{Re^{0.5}} \right]^{-1} \tag{3.166}$$

However, the parameter, *m*, can only be obtained from an experimental data source since the extent of contamination cannot be predicted theoretically. The effective diffusion coefficient that is used within this model is analogous to the Handlos–Baron equation:

$$D_{i,eff} = D_i + f_z\, E = D_i + f_z \frac{d_p\, w_t}{2048 \left(1 + \dfrac{\eta_d}{\eta_c} \right)} \tag{3.167}$$

The resulting Sh correlation for continuous mass transfer is:

$$Sh_c = \frac{2}{\sqrt{\pi}} \sqrt{\alpha} \sqrt{Pe} \tag{3.168}$$

Henschke and Pfennig (1999) recently developed a model, which is also based on the Handlos–Baron equation and introduces a constant, C_{IP}, which arises from interfacial instabilities and must also be determined experimentally. The effective diffusion coefficient reads as:

$$D_{eff}^{\sigma} = \frac{w_t\, d_p}{2048\, C_{IP} \left(1 + \frac{\eta_d}{\eta_c}\right)} \tag{3.169}$$

and can be found in the modified Fourier number:

$$Fo^t = \frac{(D_d + D_{eff}^{\sigma})\, t}{(d_p/2)^2} \tag{3.170}$$

The mass transfer with respect to an initial concentration, y_0, and its equilibrium value, y^*, is then:

$$\frac{y^* - \bar{y}(t)}{y^* - y_0} = \frac{6}{\pi^2} \sum_{n=1}^{\infty} \frac{\exp[-(n\,\pi)^2\ Fo^t]}{n^2} \tag{3.171}$$

An approximation, which is easier to handle is with

$$\frac{y^* - \bar{y}(t)}{y^* - y_0} = 1 - 6\sqrt{\frac{Fo^t}{\pi}} + 3Fo^t \quad \mathit{for}\ Fo^t \le 0.15$$

and

$$\frac{y^* - \bar{y}(t)}{y^* - y_0} = \frac{6}{\pi^2} \exp(-\pi^2 \cdot Fo^t) \quad \mathit{for}\ Fo^t > 0.15 \tag{3.172}$$

The experimental evaluation can be done either in a rising (falling) droplet apparatus (see Fig. 3.4a) for a short residence time or in a Venturi tube for long contact times [Schröter et al. (1998), Bart (2000)]. Monodispersed droplets can be produced and in the latter case captured by the counterflowing continuous phase in the conus of the Venturi tube (see Fig. 3.4b). A variation of the rising height or contact time allows an interpolation to zero height (time) and reveals the effects during drop formation or sampling [Nitsch (1989)]. The experimental data derived can be used to find m or C_{IP} in the last models.

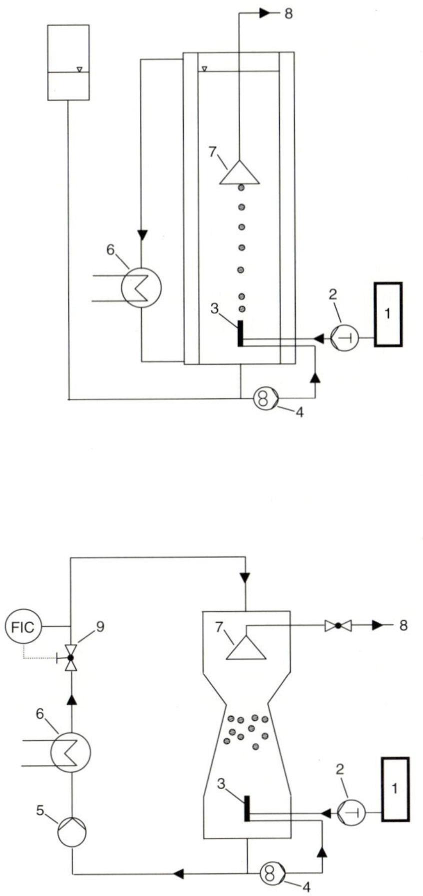

a)

b)

Fig. 3.4a,b:a Rising droplet apparatus **b** Venturi tube for droplet mass transfer experiments (1 = feed storage, 2 = melting pump, 3 = double-flow valve, 4,5 = pumps, 6 = heat exchanger, 7 = collecting funnel, 8 = stream to analysis)

3.3 Reactive Extraction Kinetics

3.3.1 Surfactants, Zeta Potential and the Initial Reaction Rate

The initial reaction rate is an appropriate tool to determine the dominating regime in a heterogeneous system. In, for example, a reactive liquid-liquid system, diffusion and reaction contribute to the macrokinetic mass transfer. The regime evaluation is only possible in a Lewis-cell type in which the two phases are separated by a flat interface and a mixing of each phase is possible without a disrupture of the planar interface (see Fig. 1.13). The mixing can be either achieved by vibration [Hancil et al. (1990)] or stirring [Nitsch (1989)], whereas the latter method is recommended. The stirring speed in both phases can be adjusted to obtain similar Reynolds' numbers and an ideally stirred vessel behaviour is achieved by specific internals. Additionally, it can be easily operated and is well documented. Fig. 3.5 shows a typical curve for the initial reaction rate in a constant interfacial cell. At low stirring speed the regime is dominated by diffusion. As the stirring speed is intensified the initial reaction rate R^0 increases due to a diminution in the interfacial diffusion resistance. At a certain stirring speed the kinetic regime is reached which is characterised by a constant R^0 despite further stirring speed intensification. Finally, the interface becomes distorted and R^0 increases again. The initial flux is the slope of the concentration versus time curves at time zero.

In a heterogeneous liquid-liquid system it is convenient to express the experimental rates in terms of molar fluxes:

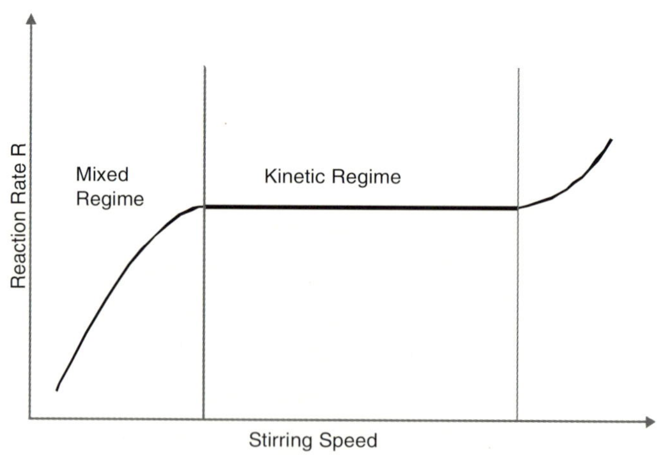

Fig. 3.5: Regimes in a stirring cell

$$N_i = -\frac{dc_i}{dt}\frac{V}{A} = R_i \frac{V}{A} \tag{3.173}$$

If c_i denotes aqueous concentrations, then V is the aqueous volume and A the interfacial area. The expression for the reaction rate, R_i, is only concentration dependent at a given temperature and is the result of kinetic investigations:

$$R = -\frac{dc_i}{dt} = f(c_1, c_2, \ldots, c_n) \tag{3.174}$$

In order to evaluate the reaction rate from experimental concentration time curves two methods are commonly used. The first one is an integration method which is numerically tedious while in the second (differential) method the rate expression can be used in its differential form [Fogler (1992)]. However, the use of rational functions is recommended [Kamenski & Dimitrov (1993)]. Such functions, e.g. of the form

$$f(t) = a\frac{t+b}{t+c} \tag{3.175}$$

show asymmetric behaviour and allow simple fitting of the data; additionally, the derivatives can be easily calculated. The initial flux is then

$$N_0 = \left.\frac{\partial f}{\partial t}\right|_{t=t_0} \frac{V}{A} \tag{3.176}$$

The reaction rate then consists of one term for the forward and one for the reverse reaction. At equilibrium both are of same magnitude and hence one kinetic constant can be replaced by the equilibrium constant. This procedure also guarantees that the kinetic data will approach the equilibrium in the limiting case of infinite time.

The rate equation (see equ. (3.174)) usually considers temperature and pressure dependence in the rate constants, the different molecular reaction steps in the final rate equation and the non-ideality of the concentrated solutions with activity coefficients. The effect of surfactants has to be taken into account in reactive extraction because, although chemically inert, they may alter the environment locally, especially when adsorbed on interfaces where the chemical reactive extraction takes place [Lin & Osseo-Asare (1986), Danesi & Chiarizia (1980), Czapla et al. (2000)].

They do not solely influence the chemical reaction but may also alter the interfacial tension and the surface mobility – the drop becomes rigid – which leads to changes in the hydrodynamics of the droplets. This in turn affects the mass transfer significantly. But the change in hydrodynamics is not the only effect that alters the mass transfer rate. The enrichment of neutral surfactants at the drop surface causes interfacial blockage resulting in difficulties for molecules to penetrate the interface. Additionally, they may influence the interfacial rheology and thus dampen Marangoni instabilities, which are different from ionic ones [Agbe & Mendes-Tatsis (2000)]. The nature of a surfactant is ambiphilic and as depicted in Fig. 3.6 has a polar hydrophilic and a non-polar hydrophobic part.

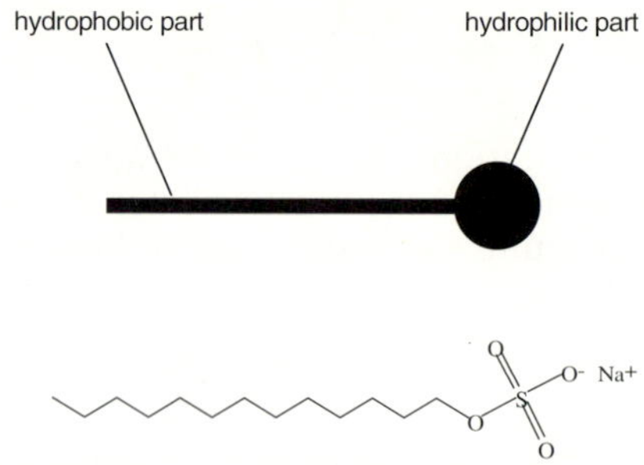

Fig. 3.6: Schematic representation of the surfactant sodium dodecyl sulfate

If mass transfer is limited by an interfacial reaction there are two main effects to be considered. The first one is *adsorption* at the interface revealing a different concentration as in the bulk phase. This is a well known fact with liquid ion exchangers due to their amphiphilic nature. The second one is a *charged barrier*, if ionic surfactants adsorb and ionic solutes have to be transferred.

The *adsorption* and aggregation behaviour of surfactants is depicted in Fig. 3.7. They increasingly adsorb at the interface while the interfacial tension decreases. At a certain concentration, the critical micelle concentration (*c.m.c.*) at the interface is fully occupied with surfactants and micelle formation starts. Above the *c.m.c.* an admixture of surfactants will only create new micelles.

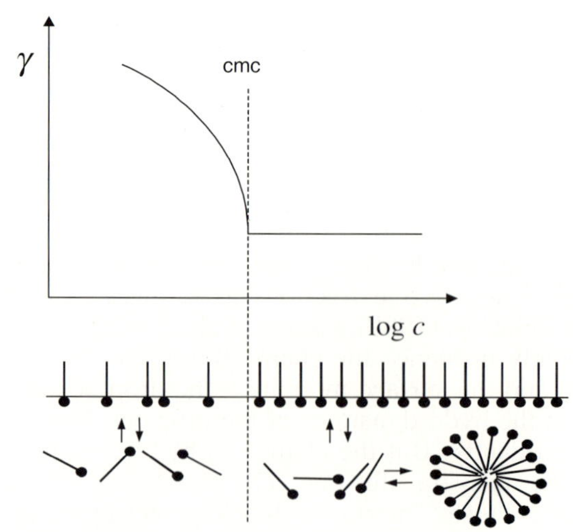

Fig. 3.7: Interfacial tension behaviour

In Fig. 3.8 the behaviour of surfactants depending on the surface aggregation is shown. At very low interfacial concentrations there is a big disorder, which is called a gas analogous phase (I). With increasing concentration or film pressure, we find a liquid expanded film (II) where several molecules aggregate to clusters. In the following developed liquid phase the interfacial area is totally covered by surfactants (III) and a slight increase leads to a liquid condensed phase (IV) of high order. At high pressure the surfactants are oriented lsimilar to a crystal structure (the so-called solid phase) (V).

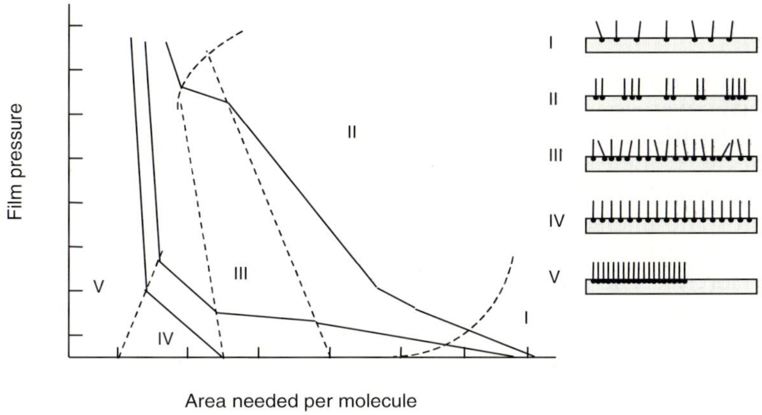

Fig. 3.8: Film pressures with surfactants

According to their functionality we know

- anionic surfactants (e.g. sodium dodecyl sulphate (*NaDdS*) sodium lauryl sulphate (*NaLS*),
- cationic surfactants (e.g. dodecyl trimethyl ammonium chloride (*DTACl*)),
- non-ionic surfactants (e.g. octyl poly (ethyleneglycolether)$_{10}$ or TRITON X-100 (*TX100*)) and
- amphoteric surfactants with zwitterionic hydrophilic groups.

The first two will dissociate in aqueous solutions or will form cationic or anionic barriers, when adsorbed at the interface. As to this, liquid cation and anion exchangers are amphiphilic by nature and may adsorb at interfaces in a liquid film. Most of the liquid ion exchangers are amphiphilic substances and tend to adsorb and concentrate at the liquid-liquid interface. This can be easily checked by interfacial tension measurements which can be used to calculate the interfacial loading according to an absorption isotherm as, for example, in a Gibbs' form:

$$\gamma = -\frac{1}{mRT}\frac{d\sigma}{d\ln c_{i,B}} \tag{3.177}$$

The interfacial coverage, γ, depends on the bulk concentration of the surfactant, $c_{i,B}$, the interfacial tension σ and m, the surfactant specific parameter which is 1 for neutral adsorbates and 2 for salt free solutions of ionic surfactants. A plot of γ versus ln c_{i^pB} reveals the saturation limit in the linear range, where the interface is totally occupied with surfactants. For many extractants this limit is achieved with solutions containing less than 0.001 mol of surfactants.

$$[RH]_{ad} = \text{const} \tag{3.178}$$

This simplification is usually implied in kinetic models. Another possibility is to state:

$$\sum [RH_i]_{ads} = \max \tag{3.179}$$

This means all amphiphilic species in the system contribute to the maximum interfacial coverage which is limited due to the interfacial area available [Roos & Bart (2000)].

However, if ionic surfactants form a *charged barrier* at the interface, this may have a considerable effect on the interfacial kinetics of ionic solutes. Due to electrostatic interactions, charge separation or polarization at the interface are typical occurrences, which can be measured as zeta-potential by electrophoresis. The electrophoretic mobility, U, can be related to the zeta potential, ζ, which is the electrokinetic potential at the plane of shear in an electrostatic polarized layer according to Fig. 3.9.

The Smoluchowski equation relates the measured mobility and the zeta potential:

$$U = \frac{\varepsilon \zeta}{\eta} \tag{3.180}$$

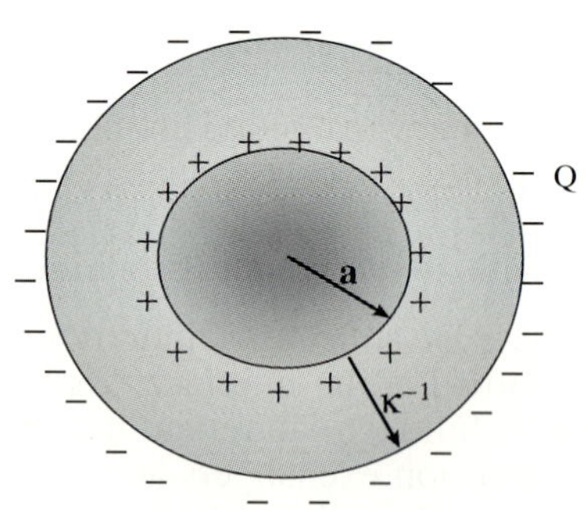

Fig. 3.9: Charge separation at a spherical particle of radius a

where ε and η are the dielectric constant and the viscosity of the continuous phase, respectively. This equation is only valid if the reciprocal Debye–Hückel parameter, κ^{-1}, is smaller than the particle radius, a, whereas κ is defined as (compare equ. (2.109)):

$$\kappa = \sqrt{\frac{e^2 N_A \sum_i c_i z_i^2}{\varepsilon k T}} \tag{3.181}$$

The general case can be described by the Henry equation [Hunter (1981)]:

$$U = \frac{2\varepsilon\zeta}{3\eta} \mathrm{f}(\kappa, a) \tag{3.182}$$

There are two limiting conditions

$\kappa\alpha < 1$

$$\mathrm{f}(\kappa, a) = 1 + \frac{(\kappa a)^2}{16} - \frac{5(\kappa a)^3}{48} - \frac{(\kappa a)^4}{96} - \left[\frac{(\kappa a)^4}{8} - \frac{(\kappa a)^6}{96}\right] \mathrm{e}^{\kappa a} \int_{\infty}^{\kappa a} \frac{\mathrm{e}^{-t}}{t} \mathrm{d}t \tag{3.183}$$

$\kappa\alpha > 1$

$$\mathrm{f}(\kappa, a) = \frac{3}{2} - \frac{9}{2\kappa a} + \frac{75}{2\kappa^2 a^2} - \frac{330}{\kappa^3 a^3} \tag{3.184}$$

The values of $f(\kappa, a)$ lie between 1 and 1.5 and approach the Debye–Hückel approximation ($\kappa \; a << 1$):

$$U = \frac{2\varepsilon\zeta}{3\eta} \tag{3.185}$$

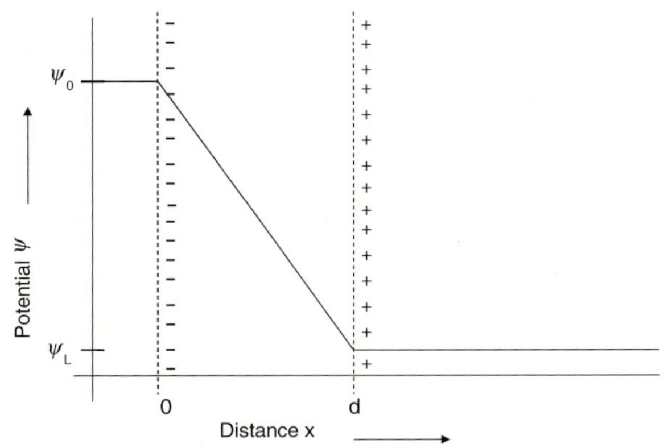

Fig. 3.10: Helmholtz double layer

From a comparison with the Smoluchowski equation it is evident that small particles move 1.5 times slower in an electrical field than large ones if the same zeta potential is applied.

In such systems two potentials occur, the zeta potential and the interfacial potential while only the latter one is of interest for reaction kinetics. Theories of electrical double layers give relations between the two. The oldest is that of a rigid double layer named after Helmholtz and is depicted in Fig. 3.10:

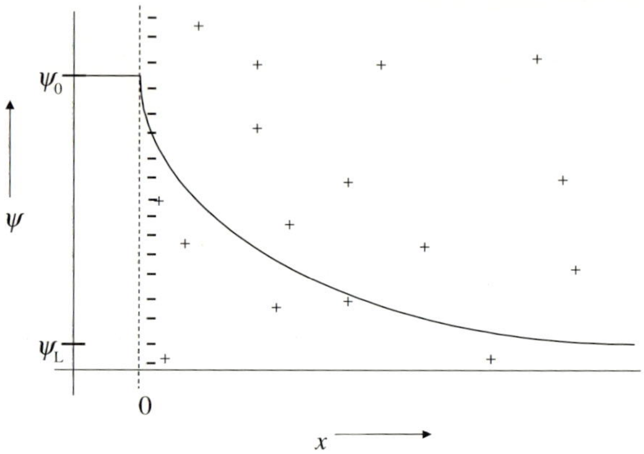

Fig. 3.11: Gouy–Chapman double layer

In Fig. 3.11 a statistical model after Gouy–Chapman is shown which accounts for the motion of ionic molecular species according to the natural Brown's thermal mobility. This model is only valid for highly diluted systems (< 1 mM), since the molecular size of the ionic species are assumed to be zero. The model of Stern is an extension of the two for real ionic solutes, with the limiting case at the inner layer described by the Helmholtz model and the outer layer according to the Gouy–Chapman model. The Poisson equation gives the relation between potential, ψ, and charge densitiy $\rho(x)$:

$$\Delta^2 \psi(x) = -\frac{\rho(x)}{\varepsilon_0 \varepsilon_r} \tag{3.186}$$

where ε_0 is the electrical field constant and ε_r the dielectrical constant. Solving this equation for the limiting case $x \to \infty$ gives the thickness of the diffusive double layer:

$$\beta = \kappa^{-1} = \sqrt{\left(\frac{\varepsilon_0 \varepsilon_r kT}{2 N_A e^2 I}\right)} \tag{3.187}$$

which is shown in Fig. 3.12. As can be seen, an increasing ionic strength compresses this layer and decreases κ^{-1}.

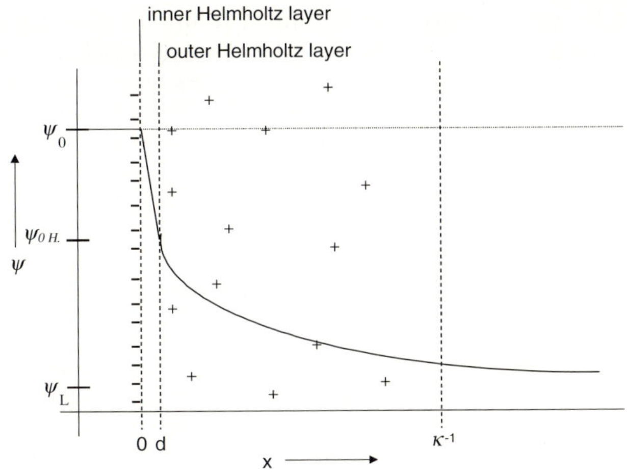

Fig. 3.12: Stern layer

The concentration of a ionic species at the interface, m_i^I, is related to the interfacial potential, ψ_0, by the Boltzmann equation:

$$m_i^I = m_i \exp\left(-\frac{z e \psi_0}{k T}\right) \tag{3.188}$$

with k, the Boltzmann constant, and e, the elementary charge. The Poisson–Boltzmann equation then reads [Hunter (1993)]:

$$\frac{\mathrm{d}^2 \psi}{\mathrm{d}x^2} = -\frac{1}{\varepsilon}\left(\sum_i z_i e c_i^o \exp\frac{-z_i e \psi}{k T}\right) \tag{3.189}$$

If the electrical energy is small compared to the thermal energy ($|z_i e \psi| <<$ $k T$) the exponent in the equation given above can be approximated with $e^{-x} = 1-x$ to give

$$\frac{\mathrm{d}^2 \psi}{\mathrm{d}x^2} = \frac{1}{\varepsilon}\left(\sum_i z_i e c_i^o - \frac{\sum_i z_i^2 e^2 c_i^o \psi}{k T}\right) \tag{3.190}$$

Because of electroneutrality in the bulk the first term is zero which gives a further simplification:

$$\frac{\mathrm{d}^2 \psi}{\mathrm{d}x^2} = \kappa^2 \psi \tag{3.191}$$

or after integrating:

$$\psi = \psi_0 \exp\left(-\kappa x\right) \tag{3.192}$$

This result can be applied only to very diluted solutions of strong monovalent electrolytes, which is the basis of the Debye–Hückel theory. With concentrated solutions of higher valent ions the linearisation of the exponent function will not hold [Schwuger (1996)]. The exact formulation of the Poisson–Boltzmann equation for a symmetric $z{:}z$ electrolyte

$$z_i = z_+ = -z_- = z \tag{3.193}$$

is

$$\tanh\left(\frac{z e \psi}{4kT}\right) = \tanh\left(\frac{z e \psi_0}{4kT}\right) \exp(-\kappa x) \tag{3.194}$$

For small ψ this equation and the approximation are equal. However, this theory allows one to predict interfacial concentrations with the knowledge of bulk concentrations, as far as the zeta potential is known. The exact position of the zeta potential is not known, but in the lack of better information it is assumed to be at the outer Helmholtz layer [Lyklema (1999)] at distance d in Fig. 3.12 and must be determined experimentally. First molecular dynamic simulations have been undertaken to evaluate its exact distance from the interface from basic principles [Lyklema et al. (1998)], but there is still a long way to go in order to apply this to industrial relevant solutions.

3.3.2 Microkinetics of Zn–D2EHPA

The microkinetics delivers the true kinetic information derived from a kinetic regime (see Fig. 3.5) without any diffusional influences which is in contrast to macrokinetics, where both diffusion and reaction are considered. In order to develop a kinetic model for a *D2EHPA* system the following basic assumptions have to be verified:

- *D2EHPA* is a dimer in kerosene like diluents
- *D2EHPA* is not soluble in water and
- the reaction site is the interface.

The last aspect has been discussed in section 3.1 and all kinds of organic acids tend to dimerise:

$$2\overline{RH} \leftrightarrow \overline{R_2H_2} \qquad K_2 = \frac{[\overline{R_2H_2}]}{[\overline{RH}]^2} \tag{3.195}$$

The distribution between water and the organic phase (indicated by a bar) is then:

$$RH \leftrightarrow \overline{RH} \qquad K_D = \frac{\overline{RH}}{RH} \tag{3.196}$$

and *D2EHPA* may then dissociate:

$$RH \leftrightarrow R^- + H^+ \qquad K_H = \frac{[R^-][H^+]}{[RH]} \tag{3.197}$$

The review of Kolarik (1982) is given in Table 3.3 and only consistent values are depicted. In aromatic diluents *D2EHPA* predominates in monomeric form. It shows an amphiphilic behaviour [Vandegrift & Horwitz (1977)] and saturates at the interface, e.g. there is a total coverage with 0.025 molar n- dodecane and 0.001 molar HNO_3 solution. Above a 0.02 molar *D2EHPA* concentration is a significant difference between activities and concentrations which leads to the recommendation by Danesi and Vandegrift (1981) to use activities rather than concentrations for the kinetic modelling. As can be seen from the values for K_D and K_H the assumptions made above are justified.

Table 3.3: *D2EHPA* in different diluents [Kolarik (1982)]

Solvent	Temperature	$-\log K_H$	$\log K_D$	$\log K_2$
n-octane	20 °C	1.25 ± 0.15	3.3 ± 0.15	4.8 ± 0.15
Chloroforme	20 °C	-	4.8 ± 0.15	4.3 ± 0.15
n-octane	20 °C	Ul'yanov (1963)	Ul'yanov (1963)	Ul'yanov (1963)
n-decane	25-50 °C	Smelov (1969)	-	-
Toluene	25 °C	Liem (1972)	Liem (1972)	Liem (1972)

In the following, various kinetic models for the *Zn-D2EHPA* system are presented. They differ mainly in the assumption as to which step is rate determining.

Model 1: The model of Ajawin et al. 1980. The overall reaction in this system is given by

$$Zn^{2+} + 1.5\overline{R_2H_2} \leftrightarrow \overline{ZnR_2(RH)} + 2H^+ \qquad K_{1,3} \tag{3.198}$$

which can be divided into the three elementary steps: adsorption of the *D2EHPA*, reaction with *Zn* and desorption of the complex formed.

$$\begin{aligned} &\overline{R_2H_2} \leftrightarrow (R_2H^-)_{ad} + H^+ && K_1 \\ &Zn^{2+} + (R_2H^-)_{ad} \xleftrightarrow{k_{v,2},k_{r,2}} (ZnR_2)_{ad} + H^+ && \text{r.d.s.} \\ &(ZnR_2)_{ad} + 0.5\overline{R_2H_2} \leftrightarrow \overline{ZnR_2RH} && K_3 \end{aligned} \tag{3.199}$$

Here the rate determining step is the formation of the solute–ion exchanger complex at the interface. The adsorption of the ion exchangers at the interface (K_1) and the replacement of the complex (K_3) due to the overall stoichiometry ($K_{1,3}$) are instantaneous steps. Based on these steps the following rate expression can be obtained.

$$-\frac{d[Zn^{2+}]}{dt} = k_{v,2}\,\frac{K_1\,[\overline{R_2H_2}][Zn^{2+}]}{[H^+]} - k_{r,2}\,\frac{[H^+][\overline{ZnR_2(RH)}]}{K_3\,[\overline{R_2H_2}]^{0.5}} \tag{3.200}$$

with $$K_{1,3} = \frac{k_{v,2}\,K_1\,K_3}{k_{r,2}}$$

which is equivalent to

$$-\frac{d[Zn^{2+}]}{dt}=\kappa_v\frac{[\overline{R_2H_2}][Zn^{2+}]}{[H^+]}-\kappa_r\frac{[H^+][\overline{ZnR_2(RH)}]}{[\overline{R_2H_2}]^{0.5}} \quad (3.201)$$

and
$$K_{13}=\frac{\kappa_v}{\kappa_r}$$

The first term in equ. (3.201) accounts for the forward reaction with a first order reaction for zinc and the *D2EHPA* dimer and −1 as the reaction order for the protons. In a double logarithmic plot the characteristic dependencies of the initial rate according to this model are given in Fig. 3.13:

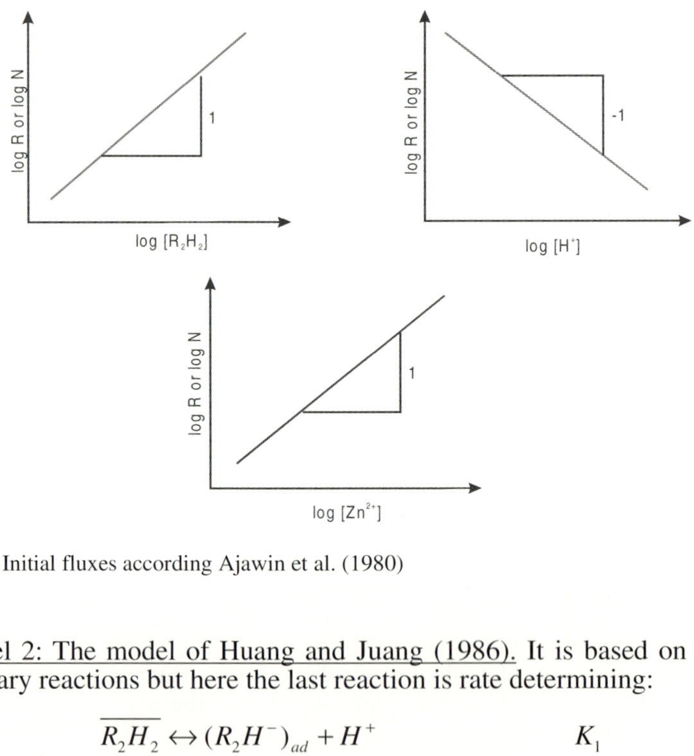

Fig. 3.13: Initial fluxes according Ajawin et al. (1980)

Model 2: The model of Huang and Juang (1986). It is based on the same elementary reactions but here the last reaction is rate determining:

$$\begin{aligned}
&\overline{R_2H_2} \leftrightarrow (R_2H^-)_{ad} + H^+ && K_1\\
&Zn^{2+} + (R_2H^-)_{ad} \leftrightarrow (ZnR_2)_{ad} + H^+ && K_2\\
&(ZnR_2)_{ad} + 0.5\overline{R_2H_2} \xleftrightarrow{k_{v,3},k_{r,3}} \overline{ZnR_2RH} && \text{r.d.s.}
\end{aligned} \quad (3.202)$$

The resulting kinetic model is then:

$$-\frac{\mathrm{d}[Zn^{2+}]}{\mathrm{d}t}=\kappa_v\frac{[\overline{R_2H_2}]^{1.5}[Zn^{2+}]}{[H^+]^2}-\kappa_r[\overline{ZnR_2(RH)}] \quad (3.203)$$

with $$K_{1,3} = \frac{\kappa_v}{\kappa_r}$$

A comparison with the previous model shows a different dependency of the initial fluxes as depicted in Fig. 3.14.

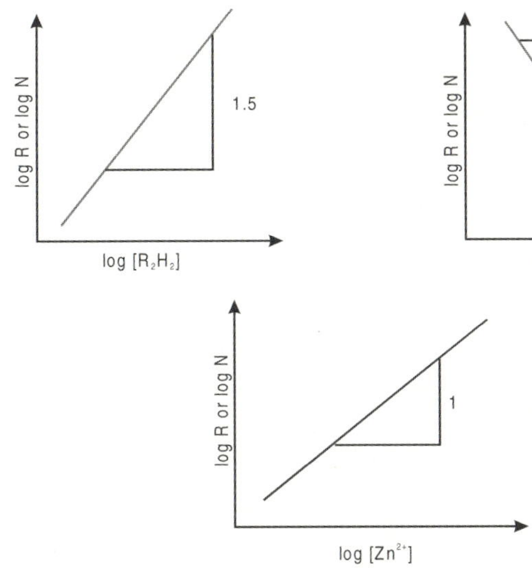

Fig. 3.14: Initial fluxes according Huang and Juang (1986)

Model 3: The model of Cianetti and Danesi (1983). In contrast to the two models presented so far the overall reaction stoichiometry is assumed to 2:1 with two consecutive reactions as the rate determining steps.

$$\begin{aligned} &Zn^{2+} + 2\overline{R_2H_2} \leftrightarrow \overline{ZnR_2(RH)_2} + 2H^+ \qquad K_{1,4} \\ &Zn^{2+} + 2(RH)_{ad} \xleftrightarrow{k_{v,1},k_{r,1}} (ZnR_2)_{ad} + 2H^+ \\ &(ZnR_2)_{ad} + 2\overline{R_2H_2} \xleftrightarrow{k_{v,2},k_{r,2}} \overline{ZnR_2(RH)_2} + 2(RH)_{ad} \end{aligned} \tag{3.204}$$

The rate law is then for the single steps:

$$-\frac{d[Zn^{2+}]}{dt} = k_{v,1}\,[Zn^{2+}][(RH)_{ad}]^2 - k_{r,1}\,[(ZnR_2)_{ad}][H^+]^2 \tag{3.205}$$

$$\begin{aligned} -\frac{d[(ZnR_2)_{ad}]}{dt} &= k_{v,2}\,[(ZnR_2)_{ad}][\overline{R_2H_2}]^2 - k_{v,1}\,[Zn^{2+}][(RH)_{ad}] \\ &\quad - k_{b,2}\,[\overline{ZnR_2(RH)_2}][(RH)_{ad}]^2 + k_{b,1}\,[(ZnR_2)_{ad}][H^+]^2 \end{aligned} \tag{3.206}$$

Considering the interface being saturated with *D2EHPA* leads to the assumption of a quasi-stationary adsorption process [Wedler (1987)]:

$$\frac{d[(ZnR_2)_{ad}]}{dt} = 0 \tag{3.207}$$

The constant ion exchanger concentration $[(RH)_{ad}]$ and the kinetic constant, k_v, can be combined to $k^{\bullet}$ which simplifies equs. (3.205) and (3.206):

$$-\frac{d[Zn^{2+}]}{dt} = k^{\bullet}_{v,1}\,[Zn^{2+}] - k_{r,1}\,[(ZnR_2)_{ad}]\,[H^+]^2 \tag{3.208}$$

$$\begin{aligned} 0 = {} & k_{v,2}\,[(ZnR_2)_{ad}]\,[\overline{R_2H_2}]^2 - k^{\bullet}_{v,1}\,[Zn^{2+}] \\ & - k^{\bullet}_{r,2}\,[\overline{ZnR_2(RH)_2}] + k_{r,1}\,[(ZnR_2)_{ad}]\,[H^+]^2 \end{aligned} \tag{3.209}$$

With equ. (3.209) $[ZnR_2]_{ad}$ can be eliminated in equ. (3.208):

$$-\frac{d[Zn^{2+}]}{dt} = \frac{k^{\bullet}_{v,1}\,[\overline{R_2H_2}]^2\,[Zn^{2+}]}{[\overline{R_2H_2}]^2 + \dfrac{k_{r,1}}{k_{v,2}}[H^+]^2} - \frac{\dfrac{k_{r,1}\,k^{\bullet}_{r,2}}{k_{v,2}}\,[H^+]^2\,[\overline{ZnR_2(RH)_2}]}{[\overline{R_2H_2}]^2 + \dfrac{k_{r,1}}{k_{v,2}}[H^+]^2} \tag{3.210}$$

with
$$K_{1,4} = \frac{k^{\bullet}_{v,1}\,k_{v,2}}{k_{r,1}\,k^{\bullet}_{r,2}}$$

or

$$-\frac{d[Zn^{2+}]}{dt} = \kappa_v\,\frac{[\overline{R_2H_2}]^2\,[Zn^{2+}]}{[\overline{R_2H_2}]^2 + C\,[H^+]^2} - \kappa_r\,\frac{[H^+]^2\,[\overline{ZnR_2(RH)_2}]}{[\overline{R_2H_2}]^2 + C\,[H^+]^2} \tag{3.211}$$

with
$$K_{1,4} = \frac{\kappa_v}{\kappa_r}\quad , \quad C = \frac{k_{r,1}}{k_{v,2}}$$

From the initial rates shown in Fig. 3.15, it can be seen that the ion exchanger and pH values show saturation areas, which is typical for all Michaelis–Menten like kinetic rates.

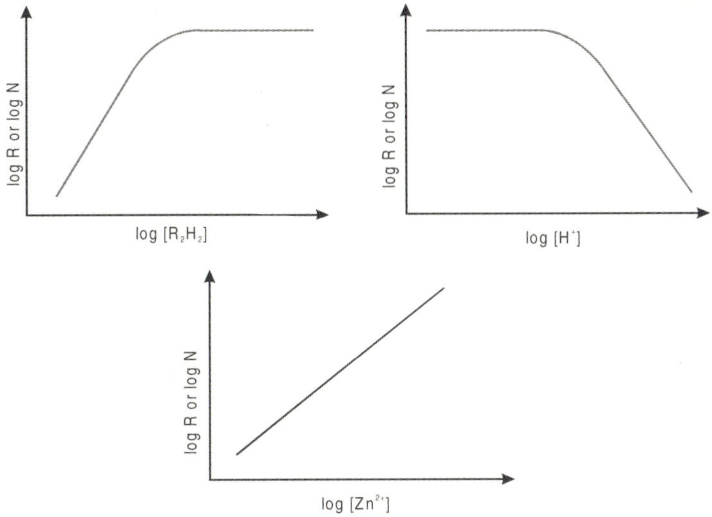

Fig. 3.15: Initial fluxes according Cianetti and Danesi (1983)

Model extensions and modifications

Model 4: Two solute–ion exchanger complexes exist. The model assumption is that the conversion of these complexes is instantaneous:

$$\overline{ZnR_2(RH)_2} \leftrightarrow \overline{ZnR_2(RH)} + 0.5\overline{R_2H_2} \qquad K_{ZnR} \tag{3.212}$$

In the final rate equation the solute–ion exchanger complex $[ZnR_2(RH)]$ in the model by Ajawin et al. (1980) is replaced by the following term which results in the model of Murthy and Ortiz (1986):

$$\frac{[\overline{ZnR_2RH}] + [\overline{ZnR_2(RH)_2}]}{1 + \dfrac{[R_2H_2]^{0.5}}{K_{ZnR}}} \text{ instead of } [\overline{ZnR_2RH}] \tag{3.213}$$

Model 5: In a similar way the complex $[ZnR_2(RH_2)]$ in the model of Cianetti and Danesi (1983) is replaced by Aparicio and Muhammed (1989) according to equ. (3.212) when considering the occurrence of two organic complexes.

$$\frac{[\overline{ZnR_2RH}] + [\overline{ZnR_2(RH)_2}]}{1 + \dfrac{K_{ZnR}}{[R_2H_2]^{0.5}}} \text{ instead of } [\overline{ZnR_2(RH)_2}] \tag{3.214}$$

Model 6: In this model the overall reaction with a 1.5:1 stoichiometric ratio [Cianetti & Danesi (1983)] can be split into a two step mechanism in which both reactions are rate determining:

$$
\begin{aligned}
&Zn^{2+} + 1.5\overline{R_2H_2} \leftrightarrow \overline{ZnR_2(RH)} + 2H^+ && K_{1,3} \\
&Zn^{2+} + 2(RH)_{ad} \xleftrightarrow{k_{v,1},k_{r,1}} (ZnR_2)_{ad} + 2H^+ && \text{r.d.s.} \quad (3.215)\\
&(ZnR_2)_{ad} + 1.5\overline{R_2H_2} \xleftrightarrow{k_{v,2},k_{r,2}} \overline{ZnR_2(RH)} + 2(RH) && \text{r.d.s.}
\end{aligned}
$$

The kinetic model that is based on this mechanism is given by:

$$
\begin{aligned}
-\frac{\mathrm{d}[Zn^{2+}]}{\mathrm{d}t} &= k_{v,1}\,[Zn^{2+}][(RH)_{ad}]^2 - k_{r,1}\,[(ZnR_2)_{ad}][H^+]^2 \\
-\frac{\mathrm{d}[(ZnR_2)_{ad}]}{\mathrm{d}t} &= k_{v,2}\,[(ZnR_2)_{ad}][\overline{R_2H_2}]^{1.5} - k_{v,1}\,[Zn^{2+}][(RH)_{ad}] \\
&\quad - k_{r,2}\,[\overline{ZnR_2(RH)}][(RH)_{ad}]^2 + k_{r,1}\,[(ZnR_2)_{ad}][H^+]^2
\end{aligned}
\tag{3.216}
$$

Again it is assumed that the interface is saturated and therefore the quasi-stationary complexation condition holds.

$$
\frac{\mathrm{d}[(ZnR_2)_{ad}]}{\mathrm{d}t} = 0 \tag{3.217}
$$

With the definition of modified kinetic constants and the constant monomeric interfacial *D2EHPA* concentration, a modified kinetic constant can be introduced:

$$
k_{v,1}^{\bullet} = k_{v,1}\,[(RH)_{ad}]^2 \text{ and } k_{v,2}^{\bullet} = k_{v,2}\,[(RH)_{ad}]^2 \tag{3.218}
$$

Elimination of $[(ZnR_2)_{ad}]$ yields a simplified kinetic model:

$$
-\frac{\mathrm{d}[Zn^{2+}]}{\mathrm{d}t} = \frac{k_{v,1}^{\bullet}\,[\overline{R_2H_2}]^{1.5}\,[Zn^{2+}]}{[\overline{R_2H_2}]^{1.5} + \dfrac{k_{r,1}}{k_{v,2}}[H^+]^2} - \frac{\dfrac{k_{r,1}\,k_{r,2}^{\bullet}}{k_{v,2}}[H^+]^2\,[\overline{ZnR_2(RH)}]}{[\overline{R_2H_2}]^{1.5} + \dfrac{k_{r,1}}{k_{v,2}}[H^+]^2}
\tag{3.219}
$$

with
$$
K_{1,3} = \frac{k_{v,1}^{\bullet}\,k_{v,2}}{k_{r,1}\,k_{r,2}^{\bullet}}
$$

Further simplification of the constants gives:

$$
R = -\frac{\mathrm{d}[Zn^{2+}]}{\mathrm{d}t} = \kappa_v\,\frac{[\overline{R_2H_2}]^{1.5}\,[Zn^{2+}]}{[\overline{R_2H_2}]^{1.5} + C\,[H^+]^2} - \kappa_r\,\frac{[H^+]^2\,[\overline{ZnR_2(RH)}]}{[\overline{R_2H_2}]^{1.5} + C\,[H^+]^2}
\tag{3.220}
$$

with $K_{1,3} = \frac{\kappa_v}{\kappa_r}$, $C = \frac{k_{r,1}}{k_{v,2}}$, $K_r = k^{\bullet}_{v,1}$, $K_r = \frac{k_{r,1}\, k^{\bullet}_{r,2}}{k_{v,2}}$

The trend of the initial flux is the same as has already been shown in Fig. 3.15.

Model 7: The elementary reactions are the same as used in model 6 but here $(RH)_{ad}$ follows the Langmuir law. Application of the quasi-stationary approach for the adsorbed zinc complex yields:

$$-\frac{\mathrm{d}[Zn^{2+}]}{\mathrm{d}t} = \frac{k^{\bullet}_{v,1}\,[\overline{R_2H_2}]^{1.5}\,[Zn^{2+}]}{[\overline{R_2H_2}]^{1.5} + \frac{k_{r,1}}{k_{v,2}}\,[H^+]^2} - \frac{\frac{k_{r,1}\,k^{\bullet}_{r,2}}{k_{v,2}}\,[H^+]^2\,[\overline{ZnR_2(RH)}]}{[\overline{R_2H_2}]^{1.5} + \frac{k_{r,1}}{k_{v,2}}\,[H^+]^2}\,[(RH)_{ad}]^2 \tag{3.221}$$

with $$K_{1,3} = \frac{k^{\bullet}_{v,1}\,k_{v,2}}{k_{r,1}\,k^{\bullet}_{r,2}}$$

The Langmuir law for the adsorbed *D2EHPA* monomer is

$$[(RH)_{ad}] = \frac{\alpha_L \frac{[\overline{RH}]}{\gamma_L}}{1 + \frac{[\overline{RH}]}{\gamma_L}} \tag{3.222}$$

and the dimerization of *D2EHPA* is assumed to be instantaneous:

$$\frac{1}{2}\overline{R_2H_2} \leftrightarrow \overline{RH} \qquad K_{0,1} \tag{3.223}$$

In combination with the law of mass action

$$K_{0,1} = \frac{[\overline{RH}]}{\sqrt{[\overline{R_2H_2}]}} \tag{3.224}$$

the final rate equation is then:

$$R = -\frac{\mathrm{d}[Zn^{2+}]}{\mathrm{d}t} = \left\{\kappa_v\,\frac{[\overline{R_2H_2}]^{1.5}\,[Zn^{2+}]}{[\overline{R_2H_2}]^{1.5} + C_1\,[H^+]^2} - \kappa_r\,\frac{[H^+]^2\,[\overline{ZnR_2(RH)}]}{[\overline{R_2H_2}]^{1.5} + C_1\,[H^+]^2}\right\} \frac{[\overline{R_2H_2}]}{(C_2 + \sqrt{[\overline{R_2H_2}]})^2} \tag{3.225}$$

with

$$K_{1,3}=\frac{\kappa_v}{\kappa_r}\quad,\ C_1=\frac{k_{r,1}}{k_{v,2}}\quad,\ C_2=\frac{\gamma_L}{K_{0,1}}\quad,\ \kappa_r=\alpha_L\,k_{r,1}\quad,\ \kappa_r=\alpha_L\,\frac{k_{r,1}\,k_{r,2}}{k_{r,2}}$$

The trend of the initial flux can be seen again in Fig. 3.15.

Model 8: The first elementary reaction in model 6 splits in two rate-determining steps which yields the following three-step mechanism:

$$\begin{aligned}
&Zn^{2+}+(RH)_{ad}\xleftrightarrow{k_{v,1},k_{r,1}}(ZnR^+)_{ad}+H^+ && \text{r.d.s.}\\
&(ZnR^+)_{ad}+(RH)_{ad}\xleftrightarrow{k_{v,2},k_{r,2}}(ZnR_2)_{ad}+H^+ && \text{r.d.s.}\\
&(ZnR_2)_{ad}+1.5\overline{R_2H_2}\xleftrightarrow{k_{v,3},k_{r,3}}\overline{ZnR_2(RH)}+2(RH)_{ad} && \text{r.d.s.}
\end{aligned}\tag{3.226}$$

Saturation of the interface with the *D2EHPA* monomer gives:

$$k^{\bullet}_{v,1}=k_{v,1}\,[(RH)_{ad}],\quad k^{\bullet}_{v,2}=k_{v,2}\,[(RH)_{ad}],\quad k^{\bullet}_{r,3}=k_{r,3}\,[(RH)_{ad}]^2 \tag{3.227}$$

and the rate equations are thus:

$$\begin{aligned}
-\frac{\mathrm{d}[Zn^{2+}]}{\mathrm{d}t}&=k^{\bullet}_{v,1}\,[Zn^{2+}]-k_{r,1}\,[(ZnR^+)_{ad}][H^+]\\
-\frac{\mathrm{d}[(ZnR^+)_{ad}]}{\mathrm{d}t}&=k^{\bullet}_{v,2}\,[(ZnR^+)_{ad}]-k^{\bullet}_{v,1}\,[Zn^{2+}]\\
&\quad-k_{r,2}\,[(ZnR_2)_{ad}][H^+]+k_{r,1}\,[(ZnR^+)_{ad}][H^+]\\
-\frac{\mathrm{d}[(ZnR_2)_{ad}]}{\mathrm{d}t}&=k_{v,3}\,[(ZnR_2)_{ad}][\overline{R_2H_2}]^{1.5}-k^{\bullet}_{v,2}\,[(ZnR^+)_{ad}]\\
&\quad-k^{\bullet}_{r,3}\,[\overline{ZnR_2(RH)}]+k_{r,2}\,[(ZnR_2)_{ad}][H^+]
\end{aligned}\tag{3.228}$$

Again the quasi-stationary approach is:

$$\frac{\mathrm{d}[(ZnR^+)_{ad}]}{\mathrm{d}t}=0\,,\quad \frac{\mathrm{d}[(ZnR_2)_{ad}]}{\mathrm{d}t}=0 \tag{3.229}$$

$[(ZnR^+)_{ad}]$ in the first rate term in equ. (3.328) is substituted by the other two which gives:

$$\begin{aligned}
R=&\frac{k^{\bullet}_{v,1}\,k^{\bullet}_{v,2}\,k_{v,3}\,[\overline{R_2H_2}]^{1.5}\,[Zn^{2+}]}{k^{\bullet}_{v,2}\,k_{v,3}\,[\overline{R_2H_2}]^{1.5}+k_{r,1}\,k_{v,3}\,[H^+][\overline{R_2H_2}]^{1.5}+k_{r,1}\,k_{r,2}\,[H^+]^2}\\
&-\frac{k_{r,1}\,k_{r,2}\,k^{\bullet}_{r,3}\,[H^+]^2\,[\overline{ZnR_2(RH)}]}{k^{\bullet}_{v,2}\,k_{v,3}\,[\overline{R_2H_2}]^{1.5}+k_{r,1}\,k_{v,3}\,[H^+][\overline{R_2H_2}]^{1.5}+k_{r,1}\,k_{r,2}\,[H^+]^2}
\end{aligned}\tag{3.230}$$

and after rearranging:

$$R = \kappa_v \frac{[\overline{R_2H_2}]^{1.5}\,[Zn^{2+}]}{[\overline{R_2H_2}]^{1.5} + C_1\,[H^+][\overline{R_2H_2}]^{1.5} + C_2\,[H^+]^2} - \kappa_r \frac{[H^+]^2\,[\overline{ZnR_2(RH)}]}{[\overline{R_2H_2}]^{1.5} + C_1\,[H^+][\overline{R_2H_2}]^{1.5} + C_2\,[H^+]^2} \tag{3.231}$$

with

$$\kappa_v = k^{\bullet}_{v,1} \quad , \quad \kappa_r \frac{k_{r,1}\,k_{r,2}\,k^{\bullet}_{r,3}}{k^{*}_{v,2}\,k_{v,3}} \, , \; K_{1,3} = \frac{\kappa_v}{\kappa_r}$$

$$C_1 = \frac{k_{r,1}}{k^{\bullet}_{v,2}} \quad , \quad C_2 = \frac{k_{r,1}\,k_{r,2}}{k^{\bullet}_{v,2}\,k_{v,3}}$$

The initial fluxes are again as in Fig. 3.15.

Model 9: Zinc is extracted by two parallel mechanisms. The adsorbed *D2EHPA* dimers dissociate instantaneously:

$$(R_2H_2)_{ad} \leftrightarrow (R_2H^-)_{ad} + H^+$$
$$K^{\bullet} = K\,[(R_2H_2)_{ad}] = [H^+][(R_2H^-)_{ad}] \tag{3.232}$$

The extraction of zinc is then:

$$Zn^{2+} + 2(R_2H_2)_{ad} \xleftrightarrow{k_{v,1},k_{r,1}} (ZnR_2)_{ad} + 2H^+ \tag{3.233}$$

$$Zn^{2+} + (R_2H^-)_{ad} \xleftrightarrow{k_{v,2},k_{r,2}} (ZnR_2)_{ad} + H^+ \tag{3.234}$$

The next step is the same as in the previous models:

$$(ZnR_2)_{ad} + 1.5\overline{R_2H_2} \xleftrightarrow{k_{v,3},k_{r,3}} \overline{ZnR_2(RH)} + (R_2H_2)_{ad} \tag{3.235}$$

All these three elementary steps (equs. (3.233)–(3.235)) are rate determining. The rate law for two parallel reactions is given below:

$$R = k^{\bullet}_{v,1}\,[Zn^{2+}] - k_{r,1}\,[(ZnR_2)_{ad}][H^+]^2 + k_{v,2}\,[Zn^{2+}][(R_2H^-)_{ad}] - k_{r,2}\,[(ZnR_2)_{ad}][H^+] \tag{3.236}$$

With the same definition of the kinetic constant $k^{\bullet}$ the reaction rate for the reaction in equ. (3.235) is then:

$$-\frac{\mathrm{d}[(ZnR_2)_{ad}]}{\mathrm{d}t} = k_{v,3}\,[(ZnR_2)_{ad}][\overline{R_2H_2}]^{1.5} - k^{\bullet}_{r,3}\,[\overline{ZnR_2(RH)}] - k^{\bullet}_{v,1}\,[Zn^{2+}] + k_{r,1}\,[(ZnR_2)_{ad}][H^+]^2 - k_{v,2}\,[Zn^{2+}][(R_2H^-)_{ad}] + k_{r,2}\,[(ZnR_2)_{ad}][H^+] \tag{3.237}$$

Analogous to the above models is the quasi-stationary assumption and equ. (3.237) can be expressed in $[(ZnR_2)_{ad}]$:

$$[(ZnR_2)_{ad}] = \frac{k^{\bullet}_{r,3}\,[\overline{ZnR_2(RH)}] + k_{v,2}\,K^{\bullet}\,\dfrac{[Zn^{2+}]}{[H^+]} + k^{\bullet}_{v,1}\,[Zn^{2+}]}{k_{v,3}\,[\overline{R_2H_2}]^{1.5} + k_{r,2}\,[H^+] + k_{r,1}\,[H^+]^2} \tag{3.238}$$

Combination with equ. (3.236) gives R:

$$\begin{aligned} R = {} & \frac{[\overline{R_2H_2}]^{1.5}\left\{k^{\bullet}_{v,1}\,k_{v,3} + \dfrac{k_{v,2}\,k_{v,3}\,K^{\bullet}}{[H^+]}\right\}}{k_{v,3}\,[\overline{R_2H_2}]^{1.5} + k_{r,2}\,[H^+] + k_{r,1}\,[H^+]^2}\,[Zn^{2+}] \\ & - \frac{[H^+]^2\left\{k_{r,1}\,k^{\bullet}_{r,3} + \dfrac{k_{r,2}\,k^{\bullet}_{r,3}}{[H^+]}\right\}}{k_{v,3}\,[\overline{R_2H_2}]^{1.5} + k_{r,2}\,[H^+] + k_{r,1}\,[H^+]^2}\,[\overline{ZnR_2(RH)}] \end{aligned} \tag{3.239}$$

However, in a system with parallel reactions one has to consider that each reaction will reach its equilibrium independently [Wedler (1987)]. Therefore the following conditions must be satisfied:

$$K_{1,3} = \frac{k^{\bullet}_{v,1}\,k_{v,3}}{k_{r,1}\,k^{\bullet}_{r,3}} \quad \text{and} \quad K_{1,3} = \frac{k_{v,2}\,k_{v,3}\,K^{\bullet}}{k_{r,2}\,k^{\bullet}_{r,3}} \tag{3.240}$$

Rearranging then yields the following reaction term with initial fluxes as in Fig. 3.15:

$$\begin{aligned} R = {} & \kappa_v\,\frac{[\overline{R_2H_2}]^{1.5}\left\{1 + \dfrac{C_1}{[H^+]}\right\}}{[\overline{R_2H_2}]^{1.5} + C_2\,[H^+] + C_3\,[H^+]^2}\,[Zn^{2+}] \\ & - \kappa_r\,\frac{[H^+]^2\left\{1 + \dfrac{C_1}{[H^+]}\right\}}{[\overline{R_2H_2}]^{1.5} + C_2\,[H^+] + C_3\,[H^+]^2}\,[\overline{ZnR_2(RH)}] \end{aligned} \tag{3.241}$$

with

$$\kappa_v = k^{\bullet}_{v,1}, \quad \kappa_r = \frac{k^{\bullet}_{v,1}}{K_{1,3}}, \quad K_{1,3} = \frac{\kappa_v}{\kappa_r}$$

$$C_1 = \frac{K^{\bullet}\,k_{v,2}}{k^{\bullet}_{v,1}}, \quad C_2 = \frac{k_{r,2}}{k_{v,3}}, \quad C_3 = \frac{k_{r,1}}{k_{v,3}}$$

Model 10 (diffusional regime): Cianetti and Danesi (1983) showed that the characteristic dependency of the initial flux in the diffusional regime can mimic a two-step interfacial reaction. The diffusional fluxes according to the two-film model are then as follows:

$$\begin{aligned} N_{Zn} &= k_w \left([Zn^{2+}]_B - [Zn^{2+}]_I \right) \\ N_H &= k_w \left([H^+]_B - [H^+]_I \right) \\ N_{R_2H_2} &= k_o \left([\overline{R_2H_2}]_I - [\overline{R_2H_2}]_B \right) \\ N_{ZnR_2(RH)_2} &= k_o \left([\overline{ZnR_2(RH)_2}]_I - [\overline{ZnR_2(RH)_2}]_B \right) \end{aligned} \tag{3.242}$$

The bootstrap relation is given by the following stoichiometry:

$$Zn^{2+} + 2\overline{R_2H_2} \leftrightarrow \overline{ZnR_2(RH)_2} + 2H^+ \qquad K_{1,4} \tag{3.243}$$

and
$$N_{Zn} = -2 \cdot N_H = -2 \cdot N_{R_2H_2} = N_{ZnR_2(RH)_2} \tag{3.244}$$

The reaction is instantaneous and at the interface the law of mass action is valid:

$$K_{1,4} = \frac{[\overline{ZnR_2(RH)_2}]_I \cdot [H^+]_I^2}{[Zn^{2+}]_I \cdot [\overline{R_2H_2}]_I^2} \tag{3.245}$$

Combinating this with the simplification [Cianetti & Danesi (1983)] that the mass transfer coefficients are equal in each phase, gives:

$$N_{Zn}\, a = C_1 \frac{[\overline{R_2H_2}]^2\, [Zn^{2+}]}{[\overline{R_2H_2}]^2 + C_3\, [H^+]^2} - C_2 \frac{[H^+]^2\, [\overline{ZnR_2(RH)_2}]}{[\overline{R_2H_2}]^2 + C_3\, [H^+]^2} \tag{3.246}$$

with
$$C_1 = k_w, \quad C_2 = \frac{k_w}{K_{1,4}}, \quad C_3 = \frac{k_w}{k_o\, K_{1,4}}$$

This rate term is mathematically identical with equ. (3.211) of model 3 and depicts the same tendencies as in Fig. 3.15, whereas the constants have a different meaning (no kinetic constants involved). This ambiguity may lead to different conclusions with regard to the reaction mechanism if the experimental data are not from a true kinetic regime (see Fig. 3.5).

Model 11 (mixed regime): The same is valid in a mixed regime [Wachter et al. (1993)] where the diffusion of zinc in the aqueous phase can be described by:

$$N_{Zn} = k_w \left([Zn^{2+}]_B - [Zn^{2+}]_I \right) \tag{3.247}$$

Additionally, the elementary reactions of the model of Cianetti and Danesi (1983) are relevant:

$$\begin{aligned} &Zn^{2+} + 2(RH)_{ad} \leftrightarrow (ZnR_2)_{ad} + 2H^+ \quad K_1 \\ &(ZnR_2)_{ad} + 2\overline{R_2H_2} \xleftrightarrow{k_{v,2},k_{r,2}} \overline{ZnR_2(RH)_2} + 2(RH)_{ad} \qquad \text{r.d.s.} \end{aligned} \tag{3.248}$$

When only the last reaction is rate determining, the following overall rate equation follows:

$$N_{Zn}\, a = C_1 \frac{[\overline{R_2H_2}]^2\,[Zn^{2+}]}{[\overline{R_2H_2}]^2 + C_3\,[H^+]^2} - C_2 \frac{[H^+]^2\,[\overline{ZnR_2(RH)_2}]}{[\overline{R_2H_2}]^2 + C_3\,[H^+]^2} \tag{3.249}$$

with $$C_1 = k_w,\quad C_2 = \frac{k_w\, k_{r,2}}{K_1\, k_{v,2}},\quad C_3 = \frac{k_w}{k_{f,2}\, K_1\, [(RH)_{ad}]^2}$$

and $$K_{1,4} = \frac{C_1}{C_2}$$

This rate equation is mathematically equivalent to equ. (3.246) for the pure diffusional regime and to equ. (3.211) for the pure kinetic regime. Care has to be taken in the model discrimination and only data from the kinetic plateau regime (see Fig. 3.5) will give the true microkinetic parameters. Thus a fit of a model with experimental data is not sufficient for discriminating the mass transfer regime.

An overview of the different models is given in Table 3.4. Qualitatively all models have a linear dependency on the zinc concentration, which cannot be used to distinguish between the models. Fig. 3.16 and Fig. 3.17 show that the initial rate has in the beginning a linear dependency with increasing *D2EHPA* or pH value and then tends to a final value. This phenomenon cannot be described with the models 1 and 2. A proper evaluation is only possible with data regression as presented in Table 3.4. However, the models 10 (diffusional regime) and 11 (mixed regime) have not been considered since only data in the highly diluted region have been taken into consideration, where the kinetic regime prevails. Since it has been proven by the equilibrium investigation that *Zn:D2EHPA* forms a 1:3 complex, model 3 (K_{14}) has not been evaluated. The back reaction constant has been substituted by the reaction equilibrium constant, which leaves only one kinetic parameter to be fitted in models 1 and 2. As can be seen from Table 3.4, these two models with only one adjustable parameter (κ_v) give the worst fit compared to the others.

Table 3.4: Microkinetic and macrokinetic models for the extraction of zinc with *D2EHPA*

Model	Reference	Comments	σ
1	Ajawin et al. (1980)	only one parameter adjustable	2.86
2	Huang & Juang (1986)	only one parameter adjustable	2.86
3	Cianetti & Danesi (1983)	K_{14}	-
4	Murthy & Perez de Ortiz (1986)	extension of model 1	-
5	Aparicio & Muhammed (1989)	extension of model 2	-
6	Wachter (1996)	like model 3 but K_{13}	1.93
7	Klocker (1996)	K_{13} + Langmuir	1.72
8	Klocker (1996)	K_{13} + three r.d.s.	1.92
9	Klocker (1996)	K_{13} + parallel mechanism	1.73
10	Cianetti & Danesi (1983)	diffusive regime	-
11	Moosbrugger (1991)	mixed regime	-

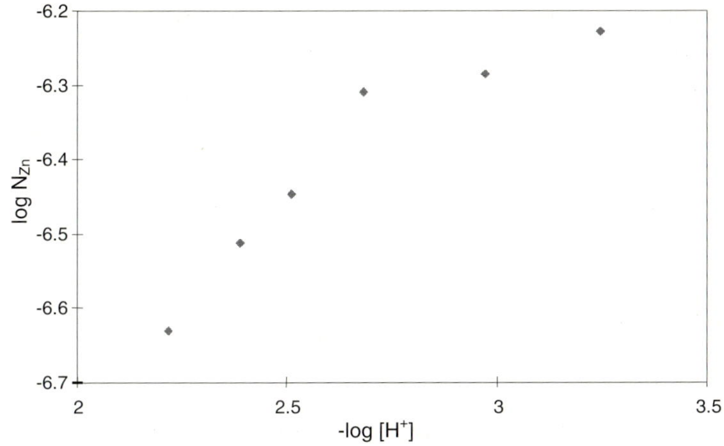

Fig. 3.16: Initial zinc flux versus pH

Model 6 has two adjustable parameters ($\kappa_{v,}$ C_1) and model 7 has three ($\kappa_{v,}$ C_1, C_2) which further improves the fit quality. In model 8 the parameter C_2 has no large influence. This parameter approaches zero if the model approximates model 7 [Klocker (1996)]. The proposed mechanism, which is often applied for three valent metal ions, is not valid for zinc. Model 9 has four adjustable parameters, but C_3 gives no improvement. With $C_3 = 0$ then $\sigma = 1.73$, which is similar to the simpler model 7.

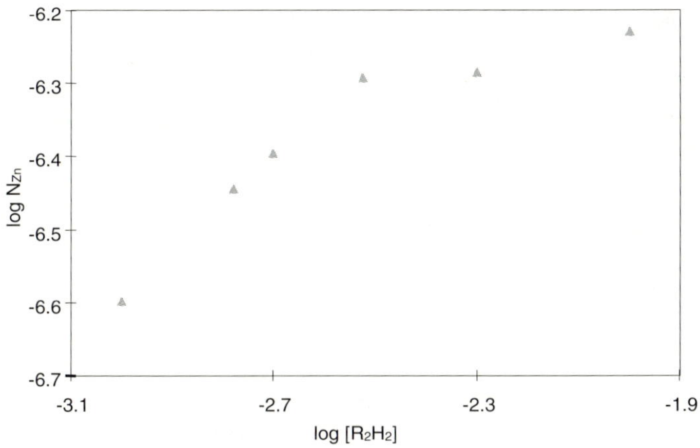

Fig. 3.17: Initial zinc flux versus log $[R_2H_2]$

As a conclusion the models 9 and 7 give similar fits. However, model 9 is of more complex nature from a chemical point of view (two parallel reactions) thus the simpler model 7 is evaluated and will be used further on. The kinetic constants for this model in n-heptane [Klocker et al. (1997)] and in isododecane [Bart & Rousselle (1999)] are given in Table 3.5 and a typical kinetic run is shown in Fig. 3.18, where the solid line is according to model 7.

Table 3.5: Estimated kinetic parameters (model 7)

Experiments at low zinc concentration	κ_v $[s^{-1}]$	κ_r $[mol^{-1/2}m^{3/2}s^{-1}]$	C_1 $[mol^{-1/2}m^{3/2}]$	C_2 $[mol^{1/2}m^{-3/2}]$	σ
Isododecane	2.3801E-04	1.0184E-04	1.2379	0.5962	1.47
n-heptane	3.2620E-04	9.0650E-05	1.1260	0.5910	1.72

For all models the dissociation equilibrium of hydrogen sulfate has to be taken into account:

$$H^+ + SO_4^{2-} \leftrightarrow HSO_4^- \tag{3.250}$$

with the equilibrium constant K_{HS}:

$$K_{HS} = \frac{\gamma_{HS}\,[HSO_4^-]}{\gamma_H\,[H^+]\,\gamma_S[SO_4^{2-}]} \tag{3.251}$$

The equilibrium constant based on molalities [Pitzer et al. (1977)] can be converted into molar units (see appendix A):

$$K_{HS}^c = K_{HS}^m\,(\rho_w)^{\sum v_i} = \frac{1}{0.0105}(\rho_w) \tag{3.252}$$

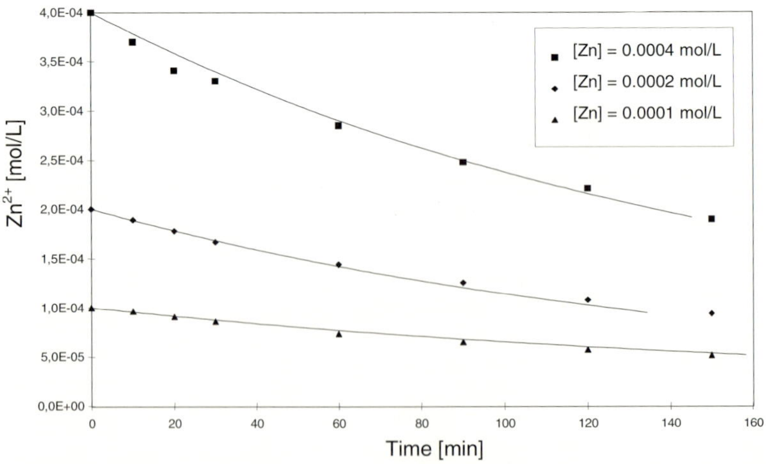

Fig. 3.18: Concentration/time diagram with varied initial zinc concentrations (solid lines according model 7)

In the above discussion we have neglected the models 10 and 11, since the data and model evaluation was in the kinetic plateau regime of the stirred cell. At higher concentrations one will enter the diffusional and mixed regime, which often prevail in industrial practice. Moreover, the hydrodynamic conditions in the planar Lewis type mass transfer cells (high *Re*-numbers in both phases) cannot be applied in a technical apparatus.

3.3.3 Macrokinetics of Zn–D2EHPA

In the following analysis the macrokinetic mass transfer across a planar interface in the mixed regime is discussed on the basis of the film model by Lewis and Whitman (1924). For free moving droplets mass transfer correlations as discussed in section 3.2.6 have to be used as will be discussed later on. The system components in the aqueous and organic phase are depicted in Fig. 3.19. The gradients in both phases are linked by the chemical reaction at the interface where zinc is extracted.

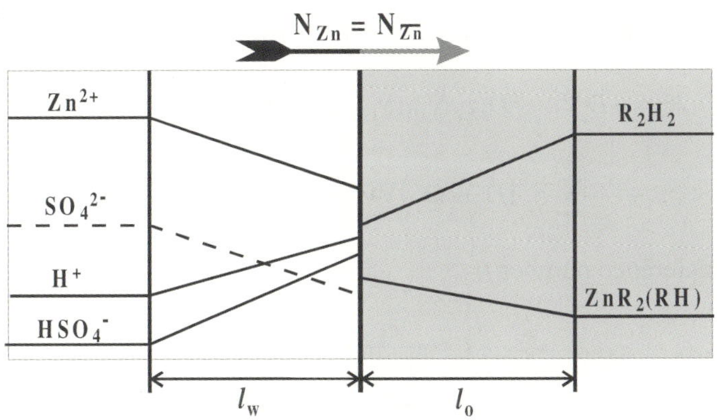

Fig. 3.19: Species gradient in the aqueous and organic film

Stoichiometric bootstraps. In the reactive system *Zn*/D2*EHPA* there is a 1:1.5 complex stoichiometry with the ion exchanger dimer R_2H_2:

$$N_{R_2H_2} = -1.5\, N_{Zn} = -1.5\, N_{\overline{Zn}} \tag{3.253}$$

The zinc flux in the aqueous phase equals the one in the organic phase:

$$N_{Zn} = N_{\overline{Zn}} \tag{3.254}$$

Zinc will exchange two protons. A part of it will react with sulfate to hydrogen sulfate and be bound there:

$$N_{Zn} = -\frac{1}{2}(N_H + N_{HS}) \tag{3.255}$$

All the other components (sulfate, water, diluent) cannot pass the interface and therefore their fluxes are zero:

$$N_S = 0 \qquad N_{H2O} = 0 \qquad N_{DIL} = 0 \tag{3.256}$$

The aqueous film. In the aqueous film we have the species water, Zn^{2+}, H^+, HSO_4^- and SO_4^{2-}. Except for the Zn^{2+} ions the other ions are linked by the hydrogen sulfate equilibrium. The basis of the film model is a time independent constant molar flux of Zn^{2+} and H^+:

$$\frac{dN_{Zn}}{dz} = 0 \quad \Rightarrow \quad N_{Zn} = \text{ const.} \tag{3.257}$$

$$\frac{dN_H}{dz} + \frac{dN_{HS}}{dz} = \frac{d(N_H + N_{HS})}{dz} = 0 \quad \Rightarrow \quad N_H + N_{HS} = \text{const.} \tag{3.258}$$

The Nernst–Planck equation for non-ideal electrolyte solutions is:

$$N_i = -D_i \nabla c_i - D_i\, c_i\, \nabla \ln\gamma_i + c_i\, u_n + \frac{t_i}{z_i\, \Im}\, i + \frac{t_i}{z_i} \sum_{j=1}^{n_I - 1} z_j \left(D_j - D_n\right) \nabla c_j + \frac{t_i}{z_i} \sum_{j=1}^{n_I} z_j\, c_j\, D_j\, \nabla \ln\gamma_j \tag{3.259}$$

with the transference number t:

$$t_i = \frac{c_i\, z_i^2\, D_i}{\sum_{j=1}^{n_I} c_j\, z_j^2\, D_j} \tag{3.260}$$

With the constraint i=0 and a zero water flux (u_n=0) equ. (3.259) reduces to:

$$N_i = -D_i\, \nabla c_i - D_i\, c_i\, \nabla \ln\gamma_i^c + \frac{t_i}{z_i} \sum_{j=1}^{n_I - 1} z_j \left(D_j - D_n\right) \nabla c_j + \frac{t_i}{z_i} \sum_{j=1}^{n_I} z_j\, c_j\, D_j\, \nabla \ln\gamma_j^c \tag{3.261}$$

In case of an one-dimensional zinc flux perpendicular to the interface there is:

$$N_{Zn} = -D_{Zn} \frac{dc_{Zn}}{dz} - D_{Zn}\, c_{Zn} \frac{d\ln\gamma_{Zn}^c}{dz} + \frac{t_{Zn}}{2} \sum_{j=1}^{3} z_j \left(D_j - D_S\right) \cdot \frac{dc_j}{dz} + \frac{t_{Zn}}{2} \sum_{j=1}^{4} z_j\, c_j\, D_j \frac{d\ln\gamma_j^c}{dz} \tag{3.262}$$

Since H^+ and HSO_4^- give the sum of H^+ exchanged with zinc, there is the following addition of fluxes:

$$N_H + N_{HS} = -D_H \frac{dc_H}{dz} - D_H\, c_H \frac{d\ln\gamma_H^c}{dz} - D_{HS} \frac{dc_{HS}}{dz} - D_{HS}\, c_{HS} \frac{d\ln\gamma_{HS}^c}{dz}$$
$$+ (t_H - t_{HS}) \sum_{j=1}^{3} z_j \left(D_j - D_S\right) \frac{dc_j}{dz} + (t_H - t_{HS}) \sum_{j=1}^{4} z_j\, c_j\, D_j \frac{d\ln\gamma_j^c}{dz} \tag{3.263}$$

The gradients of the activity coefficients can be expressed with the total differential:

$$\frac{d\,ln\gamma_i^c}{dz} = \sum_{j=1}^{4} \left\{ \left(\frac{\partial\, ln\gamma_i^c}{\partial c_j} \right)_{c_n \neq c_j} \frac{dc_j}{dz} \right\} \tag{3.264}$$

The easiest way to obtain the derivatives is solving them numerically on the basis of the Pitzer equation:

$$\left.\frac{\partial \ln\gamma_i^c}{\partial c_j}\right|_{c_j} = \frac{\ln\gamma_i^c\big|_{c_j+\Delta c_j} - \ln\gamma_i^c\big|_{c_j}}{\Delta c_j} \tag{3.265}$$

The logarithmic form of the law of mass action for the hydrogen sulfate equilibrium reads:

$$\ln K_{HS}^c + \ln\gamma_H^c + \ln c_H + \ln\gamma_S^c + \ln c_S = \ln\gamma_{HS}^c + \ln c_{HS} \tag{3.266}$$

with

$$\frac{d\,ln\,c_i}{dz} = \frac{1}{c_i}\frac{dc_i}{dz} \tag{3.267}$$

there is the relation of the gradients of H^+, SO_4^{2-} and HSO_4^-:

$$\frac{d\,ln\gamma_H^c}{dz} + \frac{1}{c_H}\frac{dc_H}{dz} + \frac{d\,ln\gamma_S^c}{dz} + \frac{1}{c_S}\frac{dc_S}{dz} = \frac{d\,ln\gamma_{HS}^c}{dz} + \frac{1}{c_{HS}}\frac{dc_{HS}}{dz} \tag{3.268}$$

One constraint is given by the electroneutrality condition:

$$2\frac{dc_{Zn}}{dz} + \frac{dc_H}{dz} - \frac{dc_{HS}}{dz} - 2\frac{dc_S}{dz} = 0 \tag{3.269}$$

However, the gradients of sulfate can be eliminated, since the electroneutrality condition must hold for the sulfate concentration

$$c_S = c_{Zn} + \frac{c_H}{2} - \frac{c_{HS}}{2} \tag{3.270}$$

The organic film. The organic phase has three components, the zinc complex $ZnR_2(RH)$, the $D2EHPA$ dimer R_2H_2, and the diluent DIL. In the

Maxwell–Stefan formulation the non-idealities are considered with the Hildebrand–Scott theory:

$$\ln\gamma_i = \frac{V_i}{RT}\left(\delta_i - \bar{\delta}\right)^2 + \ln\left(\frac{V_i}{V_m}\right) + 1 - \frac{V_i}{V_m} \tag{3.271}$$

with $$V_m = \sum_{k=1}^{n} x_k V_k$$

The molar volume V_i is in [cm^3/mol] and V_m is

$$V_m = x_{R_2H_2} V_{R_2H_2} + x_{\overline{Zn}} V_{\overline{Zn}} + x_{DIL} V_{DIL} \tag{3.272}$$

and the mean solubility parameter is:

$$\bar{\delta} = \sum_{k=1}^{3} \varphi_k \delta_k = \frac{1}{V_m} \sum_{k=1}^{3} x_k V_k \delta_k \tag{3.273}$$

For the mole fractions we need the total molar concentration of the mixture, which is assumed to be additive:

$$c_t = c_{R2H2} + c_{\overline{Zn}} + \frac{0.001 - c_{R2H2} V_{R2H2} - c_{\overline{Zn}} V_{\overline{Zn}}}{V_{DIL}} \tag{3.274}$$

The thermodynamic factor is as follows:

$$\Gamma_{ij} = \delta_{ij} + x_i \left(\frac{\partial \ln\gamma_i}{\partial x_j}\right)_{P,T,x_k \neq x_j} \tag{3.57}$$

The derivatives of the activity coefficients are calculated at constant temperature and pressure. Additionally, all the mole fractions are kept constant except for the one to be differentiated.

$$\begin{aligned}\left(\frac{\partial \ln\gamma_i}{\partial x_j}\right)_{P,T,x_i \neq x_j} &= 2\frac{V_i}{RT}\left(\delta_i - \frac{1}{V_m}\sum_{k=1}^{3} x_k V_k \delta_k\right) \\ &\left\{-A_j \sum_{k=1}^{3} x_k V_k \delta_k - \frac{1}{V_m}\left(V_j x_j - V_{DIL} x_{DIL}\right)\right\} \\ &+ V_m A_j - V_i A_j\end{aligned} \tag{3.275}$$

where A_j is defined as:

$$A_j = \left(\frac{\partial (1/V_m)}{\partial x_j}\right)_{P,T,x_i \neq x_j} = \frac{V_{DIL} - V_j}{V_m^2} \tag{3.276}$$

As mentioned earlier the matrix of the thermodynamic factors $[\Gamma]$ is assumed to be constant in the film and is calculated at the average concentration between bulk and interface

$$x_{R2H2,av} = \frac{x_{R2H2B} + x_{R2H2,I}}{2} \quad x_{\overline{Zn},av} = \frac{x_{\overline{Zn},B} + x_{\overline{Zn},I}}{2}$$
$$x_{DIL,av} = 1 - x_{R2H2,av} - x_{\overline{Zn},av} \tag{3.277}$$

The elements of the $[B]$ matrix are:

$$B_{11} = \frac{x_{R2H2,B}}{Đ_{13}} + \frac{x_{\overline{Zn},B}}{Đ_{12}} + \frac{x_{DIL,B}}{Đ_{13}}$$
$$B_{12} = -x_{R2H2,B}\left(\frac{1}{Đ_{12}} - \frac{1}{Đ_{13}}\right)$$
$$B_{21} = -x_{\overline{Zn},B}\left(\frac{1}{Đ_{12}} - \frac{1}{Đ_{23}}\right) \tag{3.278}$$
$$B_{22} = \frac{x_{R2H2,B}}{Đ_{12}} + \frac{x_{\overline{Zn},B}}{Đ_{23}} + \frac{x_{DIL,B}}{Đ_{23}}$$

The matrix of the modified mass transfer coefficients is then:

$$[k_B^*] = [k_B][\Gamma_{av}] = \frac{[B_B]^{-1}[\Gamma_{av}]}{l_o} \tag{3.279}$$

and the diffusion flux is

$$J_{R2H2,B} = -k^*_{11,B}\left(c_{R2H2,B} - c_{R2H2,I}\right) - k^*_{12,B}\left(c_{\overline{Zn},B} - c_{\overline{Zn},I}\right)$$
$$J_{\overline{Zn},B} = -k^*_{21,B}\left(c_{R2H2,B} - c_{R2H2,I}\right) - k^*_{22,B}\left(c_{\overline{Zn},B} - c_{\overline{Zn},I}\right) \tag{3.280}$$

The molar fluxes are determined due to the stoichiometric reaction bootstraps:

$$N_{R2H2} = \frac{J_{R2H,B}}{1 - 1/3\, x_{R2H2,B}} = \frac{-k^*_{11,B}\left(c_{R2H2,B} - c_{R2H2,I}\right) - k^*_{12,B}\left(c_{\overline{Zn},B} - c_{\overline{Zn},I}\right)}{1 - 1/3\, x_{R2H2,B}} \tag{3.281}$$

$$N_{\overline{Zn}} = \frac{J_{\overline{Zn},B}}{1 + 1/2\, x_{\overline{Zn},B}} = \frac{-k^*_{21,B}\left(c_{R2H2,B} - c_{R2H2,I}\right) - k^*_{22,B}\left(c_{\overline{Zn},B} - c_{\overline{Zn},I}\right)}{1 + 1/2\, x_{\overline{Zn},B}} \tag{3.282}$$

Interface. At the interface there is a zinc flux according to the interfacial reaction (model 7) which is related to the specific mass transfer area a:

$$N_{Zn} = \frac{\kappa_v}{a} \frac{\gamma^{c\;1.5}_{R2H2,I} \; c_{R2H2,I}^{1.5} \; a_{Zn,I}}{\gamma^{c\;1.5}_{R2H2,I} \; c_{R2H2,I}^{1.5} + C \, a_{H,I}^2} - \frac{\kappa_r}{a} \frac{a_{H,I}^2 \, \gamma^c_{\overline{Zn},I} \, c_{\overline{Zn},I}}{\gamma^{c\;1.5}_{R2H2,I} \; c_{R2H2,I}^{1.5} + C \, a_{H,I}^2} \tag{3.283}$$

In order to link the film side mass transfer with the interfacial reaction we have to convert both types of activity coefficients, the molal Pitzer coefficients for the aqueous phase and the mole fraction based Hildebrand–Scott coefficients for the organic phase to the molar frame (see also appendix B). However, for the latter ones we need the activity coefficients at infinite dilution, which are defined as:

$$\ln \gamma^\infty_{R2H2} = \frac{V_{R2H2}}{RT} \left(\delta_{R2H2} - \delta_{DIL}\right)^2 + \ln\left(\frac{V_{R2H2}}{V_{DIL}}\right) + 1 - \frac{V_{R2H2}}{V_{DIL}} \tag{3.284}$$

$$\ln \gamma^\infty_{\overline{Zn}} = \frac{V_{\overline{Zn}}}{RT} \left(\delta_{\overline{Zn}} - \delta_{DIL}\right)^2 + \ln\left(\frac{V_{\overline{Zn}}}{V_{DIL}}\right) + 1 - \frac{V_{\overline{Zn}}}{V_{DIL}} \tag{3.285}$$

and thus

$$\gamma^c_{R2H2,I} = \gamma_{R2H2,I} \left(\frac{x_{R2H2,I}}{c_{R2H2,I}}\right) \left(\frac{1/V_{DIL}}{\gamma^\infty_{R2H2}}\right) = \frac{1}{c_t \, V_{DIL}} \left(\frac{\gamma_{R2H2,I}}{\gamma^\infty_{R2H2}}\right)$$
$$\gamma^c_{\overline{Zn},I} = \gamma_{\overline{Zn},I} \left(\frac{x_{\overline{Zn},I}}{c_{\overline{Zn},I}}\right) \left(\frac{1/V_{DIL}}{\gamma^\infty_{\overline{Zn}}}\right) = \frac{1}{c_t \, V_{DIL}} \left(\frac{\gamma_{\overline{Zn},I}}{\gamma^\infty_{\overline{Zn}}}\right) \tag{3.286}$$

After eliminating $\gamma_{R_2H_{2,I}}$ and $\gamma_{Zn,I}$ by the Hildebrandt–Scott equation one yields:

$$\gamma^c_{R2H2,I} = \frac{\exp\left(\frac{V_{R2H2}}{RT}\left(\delta_{R2H2} - \overline{\delta}\right)^2 + \ln\left(\frac{V_{R2H2}}{V_m}\right) + 1 - \frac{V_{R2H2}}{V_m}\right)}{c_t \, V_{DIL} \, \gamma^\infty_{R2H2}} \tag{3.287}$$

$$\gamma^c_{\overline{Zn},I} = \frac{\exp\left(\frac{V_{\overline{Zn}}}{RT}\left(\delta_{\overline{Zn}} - \overline{\delta}\right)^2 + \ln\left(\frac{V_{\overline{Zn}}}{V_m}\right) + 1 - \frac{V_{\overline{Zn}}}{V_m}\right)}{c_t \, V_{DIL} \, \gamma^\infty_{\overline{Zn}}} \tag{3.288}$$

The mean solubility parameter $\overline{\delta}$ and V_m are:

$$\overline{\delta} = \sum_{k=1}^{3} \varphi_k \, \delta_k = \frac{1}{V_m} \left\{ V_{DIL} \, \delta_{DIL} + \sum_{k=1}^{2} \frac{c_k}{c_t} \left(V_k \, \delta_k - V_{DIL} \, \delta_{DIL}\right) \right\} \tag{3.289}$$

$$V_m = \frac{c_{R2H2,I}}{c_t}\left(V_{R2H2} - V_{DIL}\right) + \frac{c_{\overline{Zn},I}}{c_t}\left(V_{\overline{Zn}} - V_{DIL}\right) + V_{DIL} \tag{3.290}$$

The last two equations have been derived from equ. (3.272), where the mole fraction of *D2EHPA* and $ZnR_2(RH)$ are replaced by c_i/c_t. Additionally the summation condition for all species is

$$x_{DIL,I} = 1 - x_{R2H2,I} - x_{\overline{Zn,I}} \tag{3.291}$$

The closing condition is then

$$N_{R_2H_2} = -1.5\, N_{\overline{Zn}} \tag{3.292}$$

respectively $$N_{R_2H_2} = -1.5\, N_{Zn} = -1.5\, N_{\overline{Zn}} \tag{3.293}$$

Finally, we obtain an equation system with 14 unknown variables,

- four aqueous concentration gradients (zinc, protons, sulfate, hydrogen sulfate),
- four aqueous activity coefficents gradients (same species as above),
- two interfacial concentrations (zinc-*D2EHPA* complex, *D2EHPA* dimer),
- two interfacial activity coefficients (same species as above),
- two molar fluxes (zinc-*D2EHPA* flux, *D2EHPA* dimer flux)

which can be estimated when solving six non-linear equations for the organic phase (equs. (3.292), (3.288), (3.287), (3.283), (3.282), (3.281)) and four for the aqueous phase (equs. (3.269), (3.268), (3.263), (3.262) and four times (3.264)). When using the reaction kinetics parameters from Table 3.5, the agreement factor is 2.08 if the aqueous film thickness is assumed to be zero ($l_w = 0$) and the diffusion resistance is only considered in the organic phase with n-heptane as diluent ($l_o = 0.102 \pm 0.01$ mm). An alternative use of the penetration theory instead of the film model reveals no major improvement [Klocker (1996)] to describe the mixed regime. Klocker (1996) could show that the matrix of the mass transfer is dominated by the main diagonal elements and have been just recently determined by the Taylor dispersion technique in the group of Hampe [Mack (2000)]. Usually, the off-diagonal elements contribute to about 10 %; but in some cases they are even larger than the main elements. A pseudo-binary treatment as follows is not recommended:

$$[k_B] = \frac{1}{l_o}\begin{bmatrix} D^0_{R_2H_2,DIL} & 0 \\ 0 & D^0_{\overline{Zn,DIL}} \end{bmatrix} \tag{3.294}$$

In the Figs. 3.20, 3.21 and 3.22 the concentration–time curves for a wide span in aqueous zinc and pH and organic *D2EHPA* concentrations are given according to the model evaluated. As can be seen, the mixed regime model describes the experimental data satisfactorily.

The mass transfer experiments at planar interfaces can be concluded as follows:

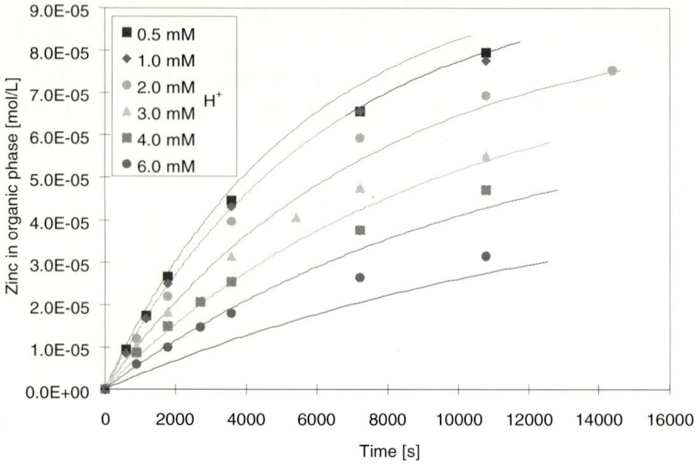

Fig. 3.20: Concentration-time diagram with varied initial zinc concentrations ($[RH]_o$ = 0.01 M, $[H^+]_o \equiv$ 1mM)

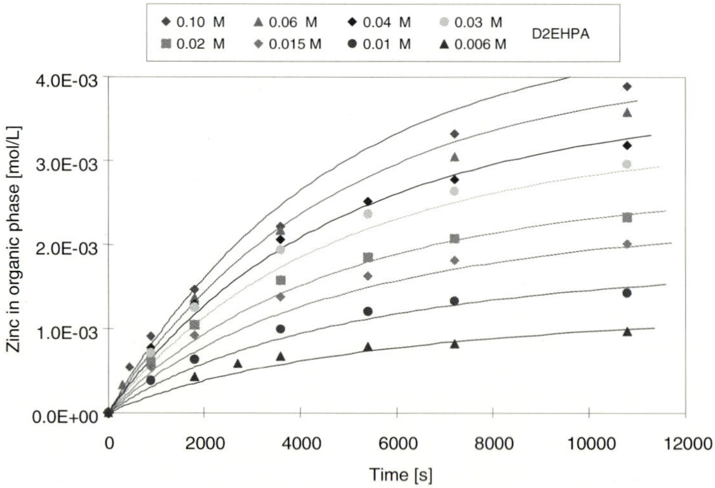

Fig. 3.21: Concentration–time diagram with varied initial pH value ($[Zn]_o$ = 0.1 mM, $[H^+]_o$ = 0.01mM)

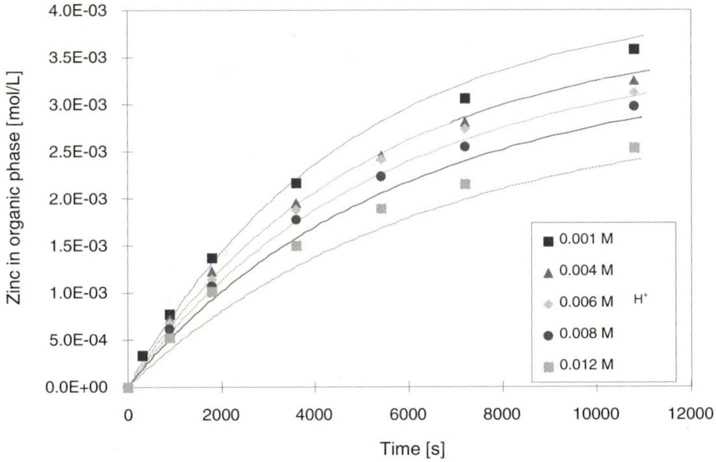

Fig. 3.22: Concentration–time diagram with varied initial *D2EHPA* ($[Zn]_o$=5 mM, $[H^+]_o$= 1 mM)

- the film model without a high flux correction is sufficient
- we have two adjustable parameters, the interfacial diffusional film thicknesses (l_o, l_w) to correlate the results while the aqueous one can be neglected.

In a technical apparatus we will always find droplets which imply a spherical geometry with mobile and immobile surfaces. Mörters and Bart (2000b) showed with LIF[1]-measurements that the dominating mass transfer prevails inside a droplet when using rhodamine as a fluorescence active indicator (see Fig. 3.23). However, for such measurements the droplet has to be fixed on a capillary which then implies an immobile surface (that is a rigid droplet behaviour).

The calculated mean diffusion coefficient from the experimental profiles is 3.71×10^{-10} m^2/s which is in good accordance with the prediction of Wilke and Chang (see equ. (3.123)) with 3.2×10^{-10} m^2/s. Free moving droplet experiments in a Venturi-tube or in a rising droplet apparatus (see Fig. 3.4) will reveal the conditions similar to extraction columns or mixers–settlers. The experimentally detected concentration profiles can be correlated with the following formula:

$$\frac{\partial C(t)}{\partial t} = -k_y\, a\,(C(t) - C^*) \qquad (3.295)$$

where a is the specific interfacial area, $C(t)$ is the concentration in the organic phase and the aqueous phase concentration C^* remains nearly constant due to the extreme phase ratio. An integration yields:

[1] LIF: laser induced fluorescence

$$C(t) = C^*(1 - \exp[-k_y\, a\,(t - t_0)] \tag{3.296}$$

t_0 is the time for droplet formation at the nozzle and coalescence in the sampling tube. It can be eliminated when interpolating all the experiments towards zero rising height or contact time. k_y is a global mass transfer coefficient which can then be compared with model predictions.

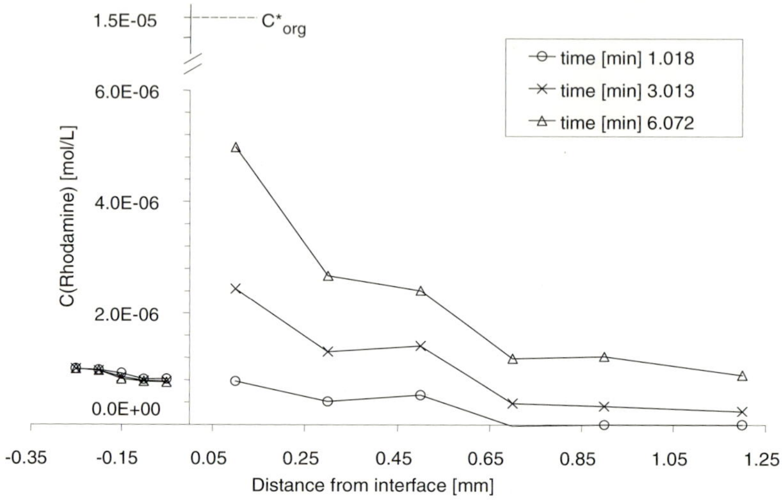

Fig. 3.23: Concentration profiles inside (right) and outside (left) of a fixed droplet in a stagnant environment (C^*_{org} is the equilibrium solubility)

In the following analysis the microkinetic model parameters for *D2EHPA* in isododecane were used (see Table 3.5); aqueous phase diffusional resistances were neglected throughout the analysis. The diffusion in a rigid organic droplet was calculated with the Maxwell–Stefan model where species 1 is $(D2EHPA)_2$, species 2 is $Zn\,(D2EHPA)_3$ and species 3 is isododecane. The necessary bootstrap relation is given assuming isododecane to be insoluble in water ($N_3 = 0$). The molecular species diffusivities were calculated after Wilke–Chang while the cross-coefficients were neglected. The driving force is then:

$$d_i = x_i \frac{\mathrm{d}\, ln\, a_i}{\mathrm{d}r} \tag{3.297}$$

and is linearised (see Fig. 3.24):

$$\frac{\mathrm{d}\, ln\, a_i}{\mathrm{d}r} = \frac{1}{\overline{a}_i}\frac{\mathrm{d}a_i}{\mathrm{d}r} = \frac{1}{\Delta r}\frac{\Delta a_i}{\overline{a}_i} \tag{3.298}$$

The activities between two adjacent layers are given by [Roos (2000)]:

$$\frac{\Delta a_i}{\bar{a}_i} = \frac{\gamma_{i,a}\, C_{i,a} - \gamma_{i,b}\, C_{i,b}}{0.5\,(\gamma_{i,a}\, C_{i,a} + \gamma_{i,b}\, C_{i,b})} \tag{3.299}$$

The diffusional mass transfer from layer a to b is then:

$$N_i = D_i\, x_i\, C_{tot} \left\{ \frac{\gamma_{i,a}\, C_{i,a} - \gamma_{i,b}\, C_{i,b}}{0.5(\gamma_{i,a}\, C_{i,a} + \gamma_{i,b}\, C_{i,b})} \right\} \tag{3.300}$$

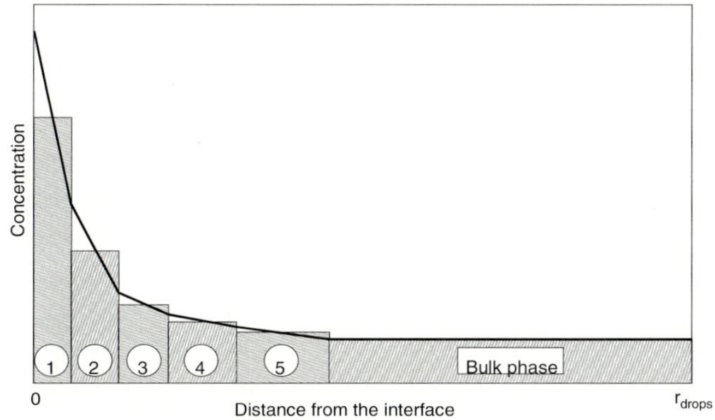

Fig. 3.24: Integration domains inside a droplet

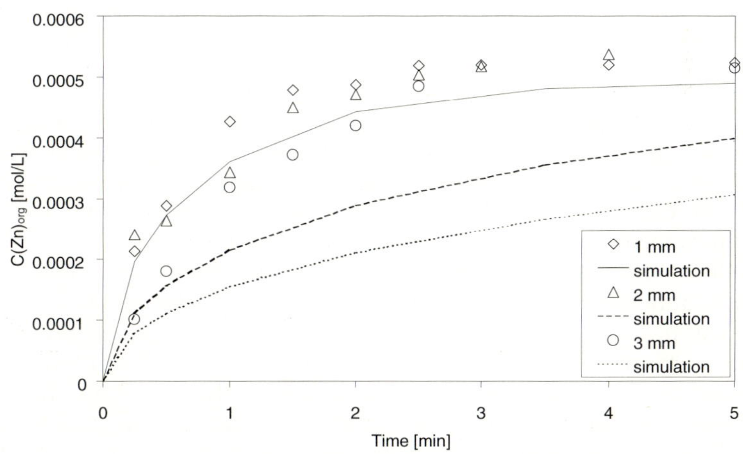

Fig. 3.25: Calculated profiles due to the molecular diffusion model combined with the interfacial reaction model (parameter = r)

The model predictions, when assuming a simultaneous interfacial reaction combined with molecular diffusion inside the droplet, are shown in Fig. 3.25. As can be seen, the predicted profiles give a lower mass flux than the experimental ones. This is in contrast to Fig. 3.26, where diffusional influences are neglected and only the interfacial reaction dominates the mass transfer.

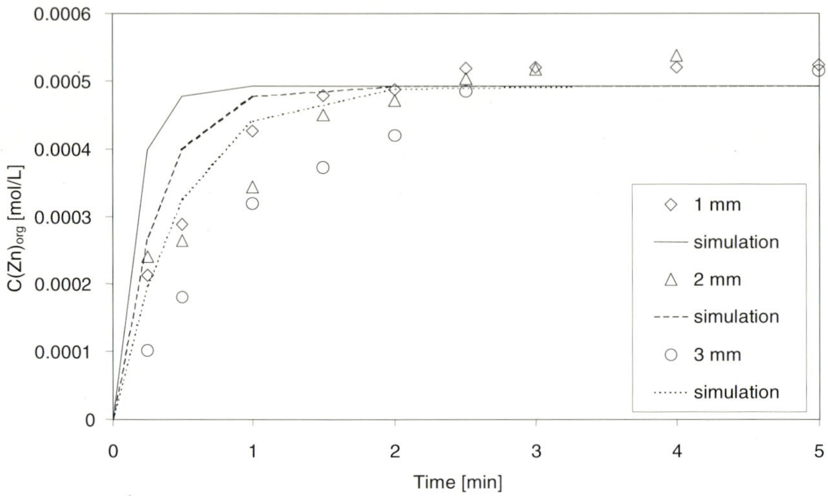

Fig. 3.26: Calculated profiles with the interfacial reaction model (parameter = *r*)

However, since experiments with fixed droplets in a capillary (rigid droplet with immobile surface) can be well described with the molecular diffusion model (see Fig. 3.27), there is the indication that considering the droplet hydrodynamics and an effective diffusion coefficient will lead to improved results. In Fig. 3.28 we see the model predictions with reaction combined with the stagnant cap model (see equ. (3.167)) where f_z = 0.12 and respectively in Fig. 3.29 we see the results of the model of Henschke and Pfennig (see equ. (3.169)) with $CIP = 9.1$. Fig. 3.30 shows the model predictions due to Steiner (1986) which need no system specific adjustable parameter.

The total experimental domain in which there are droplet sizes of 1, 2 and 3 mm, concentrations of 5, 20, 60 and 100 μmol/L *D2EHPA* and 0.1, 5 and 10 μmol/L *Zn* at pH = 3 is depicted in Fig. 3.31 according to equ. (3.296). It gives the model predictions according to the model of Steiner (1986) which gives the best parameter ($\sigma = 5.93$). The predictions only suffer in the higher concentration domain which is a problem of all models investigated here. The pure kinetic model (see Fig. 3.26) gives the worst results with $\sigma = 20.09$ and molecular diffusion combined with reaction (see Fig. 3.25) is next with $\sigma = 8.25$. All the other models are of the same prediction quality. The stagnant cap model combined with reaction (see Fig. 3.28) gives $\sigma = 7.48$ and the model of Henschke and Pfennig in combination with reaction (see Fig.

3.29) gives $\sigma = 6.67$. The two last mentioned models need one adjustable parameter (f_z and C_{IP} respectively), which is in contrast to the model of Steiner combined with reaction (see Fig. 3.30) which needs no adjustable parameter. However, since the regression parameters (besides the pure kinetic model) are very similar it is difficult to decide which model is really true. Since the models suffer in the higher concentration domain additional work on further

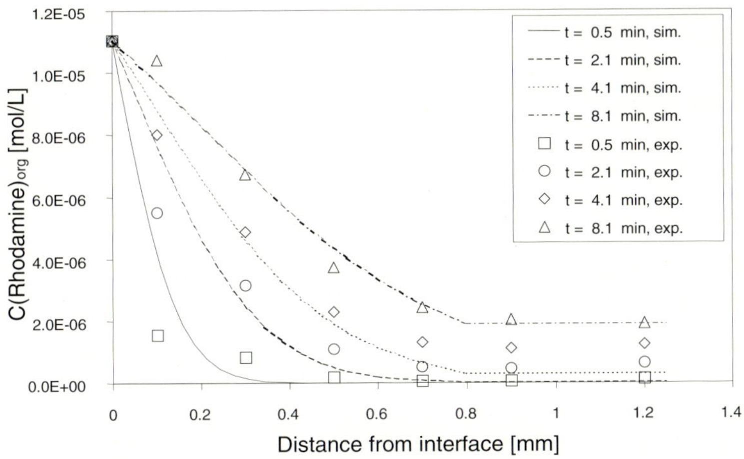

Fig. 3.27: Rhodamine transport with the molecular diffusion model

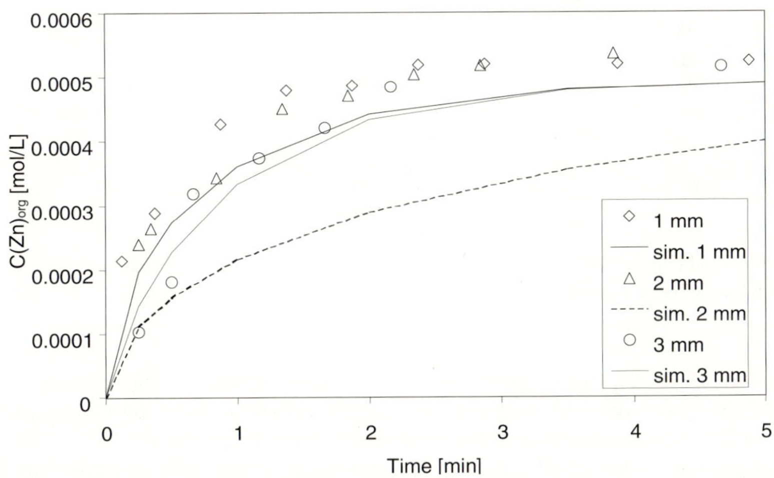

Fig. 3.28: Calculated profiles with the stagnant cap model combined with the interfacial reaction model (parameter = r)

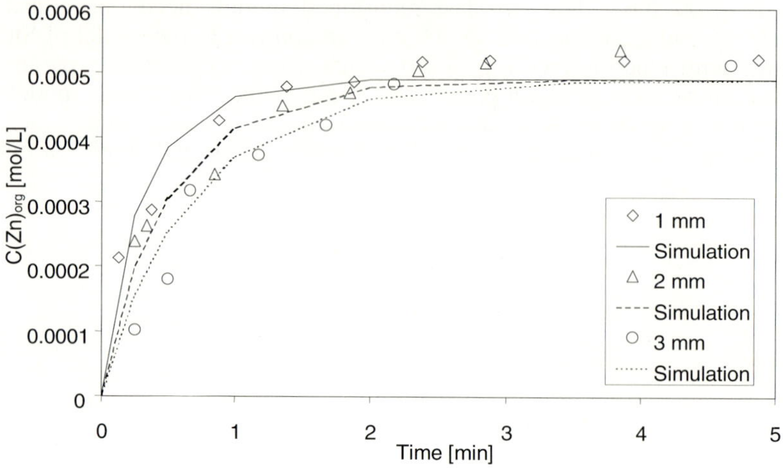

Fig. 3.29: Calculated profiles with the effective mass transfer model of Henschke and Pfennig combined with the interfacial model (parameter = r)

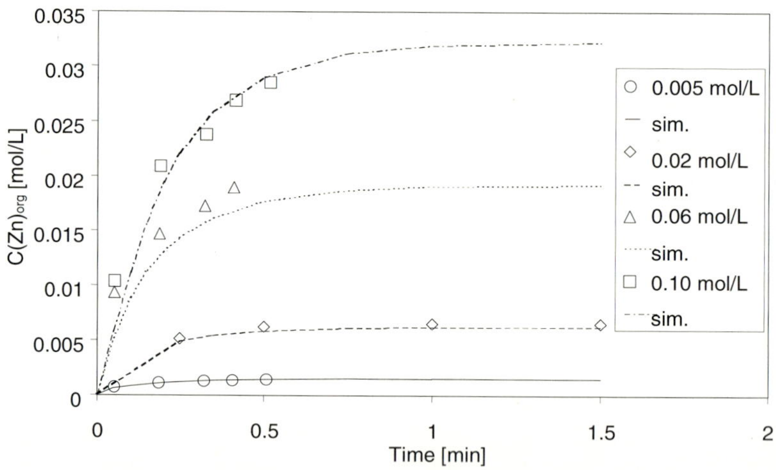

Fig. 3.30: Calculated profiles with the effective mass transfer model of Steiner combined with the molecular diffusion model

improvements is required [Mörters (2000)]. The picture is more complicated in columns, since the correlations for single droplets have to be adapted in regard to the dropled swarm behaviour in a certain geometry. As so far, the mass transfer with the *Zn/D2EHPA* system in an extraction column has been investigated by the group of Slater and is presented elsewhere [Corsi et al. (1998), Veglio & Slater (1996)].

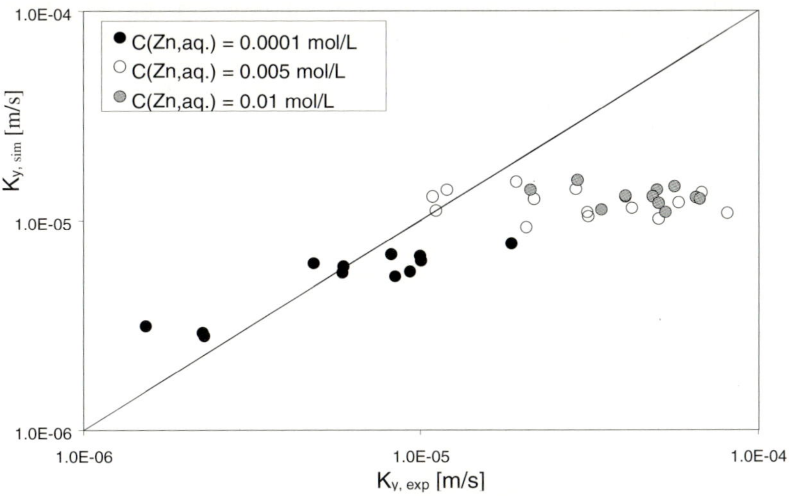

Fig. 3.31: Comparison of experimental and predicted mass transfer coefficients with the mass transfer model of Steiner combined with the interfacial reaction model

3.3.4 Microkinetics of Acid Extraction with TOA

The development of a kinetic model with anion exchange systems and its experimental verification is quite similar to the previous chapter with the cation exchange systems. In an extension to the models that have been already derived we will discuss the influence of adsorption layers on the interfacial transport and the effect of a competitive extraction of a second solute. Such unspecific co-extraction is often the case with liquid anion exchangers since their selectivities are usually rather poor. It will be considered with the LAP[2]-model which balances all the interfacially adsorbed species at saturation.

The presence of interfacial active substances is of high technical relevance for reactive extraction processes due to their favoured enrichment at the interface. When ionic surfactants are involved there will be a charge separation at the interface, which can be experimentally verified by zeta potential measurements [Lin & Osseo-Assare (1986), Hunter (1993)]. The coverage and therefore the amount of surfactants per unit area adsorbed at the interface at a given temperature can be calculated on the basis of the Gibbs equation of adsorption (see equ. (3.177)). In the following comparison we will discuss the influence of non-ionic and ionic adsorption layers on mass transfer for an equal interfacial coverage of 0.75 $\mu mol/m^2$ of surfactant. In the system *TOA*/water the non-ionic surfactant Triton X-100 (*TX100*) aggregates to micelles at a lower bulk concentration (one decimal power) than the ionic ones (see Fig. 3.32). The anionic ones (sodium lauryl sulfate, *NaLS*, and sodium dodecane sulfonate, *NaDdS*) aggregate at higher values than *TX100* and at

[2] LAP: limited number of adsorption places

lower than the cationic ones (dodecyltrimethylammonium chloride, *DTACl*). The depicted lines are limited by the critical micelle concentration (c.m.c.) and consequently the influence of the amphiphiles on reactive mass transfer is solely caused be the charge density at the interface [Czapla & Bart (2000a,b)].

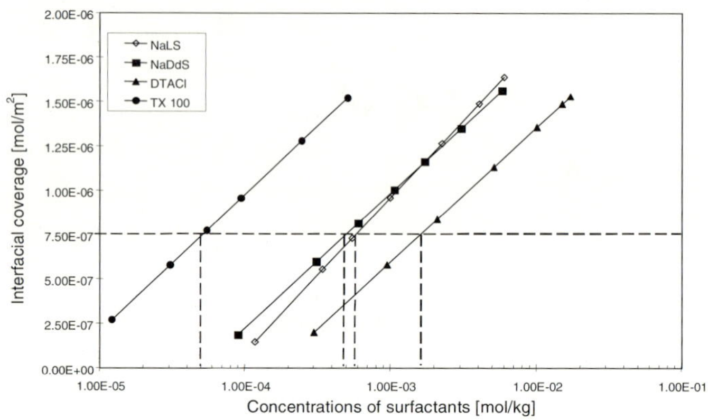

Fig. 3.32: Interfacial coverage of surfactants per unit area calculated using the Gibbs equation (10 % *TOA* in toluene, 298 K, c.m.c. for all surfactants > 1.5 μmol/m^2)

Figure 3.33 shows the influence on the zeta potential in a surfactant free (10 % *TOA*) and surfactant laden interface within a wide pH range. The non-ionic surfactant does not affect the zeta potential and for *DTACl* the zeta potential is

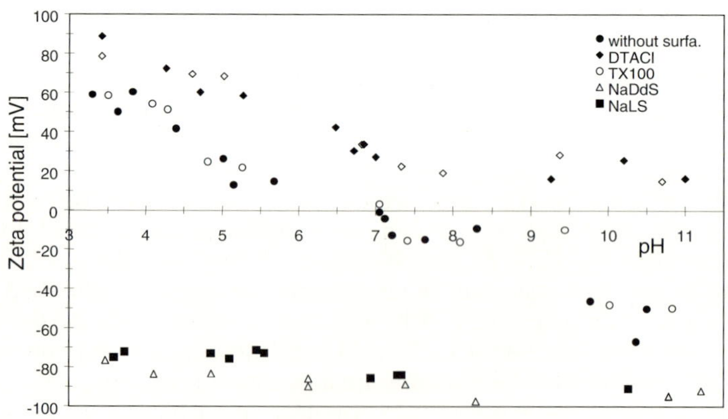

Fig. 3.33: Zeta potential–pH curves for 10 % TOA in toluene in the presence of surfactants at an interfacial covering of 0.75 μmol/m^2

positive while for the anionic surfactants (*NaDdS*, *NaLS*) the zeta potential becomes negative in about the same way.

The influence of these surfactants on acetic acid extraction in a mass transfer cell (see Fig. 1.13) is depicted in Fig. 3.34. It shows the significant plateau regions for the calculated initial mass transfer of the surfactant-free system in comparison to the amphiphilic ones used. The induced charge at the interface which is caused by a hygroscopic–nonhygroscopic orientated coverage of surfactants at the interface, has a strong influence on the initial flux. The anionic active layers decrease the flux while it is increased by cationic surfactants and independent from non-ionic surfactants at an equal coverage and charge density. Furthermore, there is no difference between the two anionic surfactants *NaDdS* and *NaLS* which means that the change in flux is determined by the induced charge and not by the adsorbed substance. This is a very important fact because the adsorbate matrix and therefore also the charge at the interface has a significant effect on the interfacial kinetics.

Figure 3.35c represents the case of a cationic-charged interface where the anionic extractable aqueous species (acetate anion) is preferably accumulated in the resulting double layer due to electrostatic interactions. Therefore, the corresponding hydrogen protons have free access to enhanced protonation of the interfacially adsorbed *TOA* molecules. This and the high acetate concentration in the close vicinity are the reasons for an increase of the extraction rate. This corresponds with the upper values in Fig. 3.34. The opposite effect is shown in Fig. 3.35a where the protons are bound in the double layer and make the protonation of the adsorbed *TOA* molecule more difficult. Additionally, the negative charge of the adsorption layer repulses the acetate anion which leads to a decrease of the extraction rate as depicted in the lower values in Fig. 3.34. The neutral case is shown in Fig. 3.35b, the equal distribution of hydrogen protons and acetate anions at the interface leads to a balanced extraction rate which is comparable to the surfactant-free system (see the middle values in Fig. 3.35). So the presence of a loaded organic/aqueous

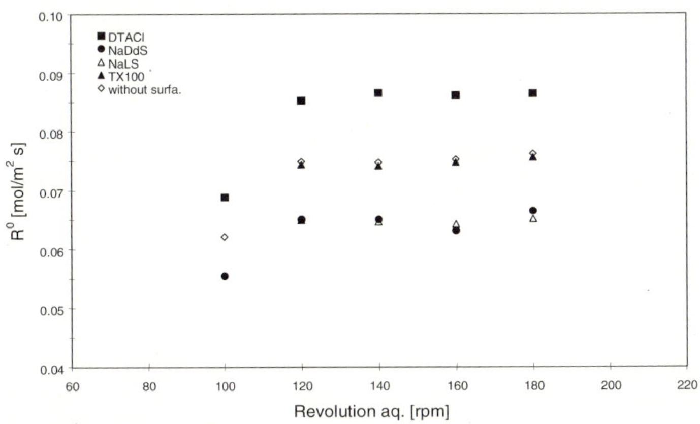

Fig. 3.34: Influence of surfactants on the mass transfer in the reaction limited regime, γ=7.5×10^{-7} mol/m², 0.2 wt.% *HAc* / 10 wt.% *TOA* in toluene (298 K)

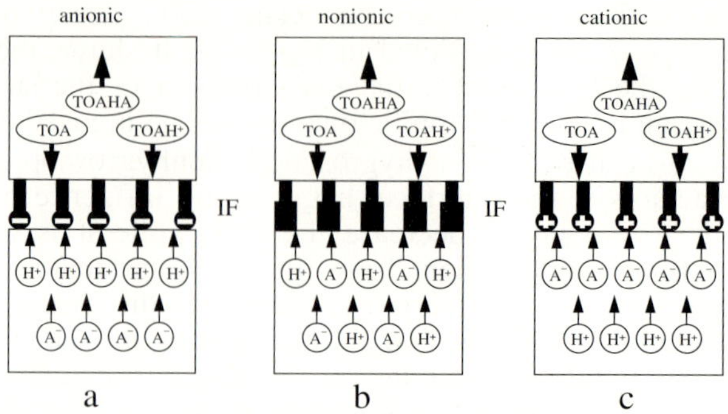

Fig. 3.35: Schematic representation of a surfactant laden interface (IF)

interface changes the interfacial concentration of the extractable aqueous species and thus the extraction rate.

On the basis of the experimental results a kinetic model and its electrostatic extension were developed to describe the interfacial reactions and the mass transfer mechanism, when using a thermodynamic consistent activity coefficient model for the aqueous and organic phase. The aqueous activity coefficients were calculated with the Pitzer theory and the Hildebrand and Scott solubility parameters were used for the calculation of the organic activity coefficients likewise in the *Zn-D2EHPA* system. The modelling of the phase equilibria was executed neglecting the co-extraction of water during the complex formation of the acetic acid in the organic phase. Nevertheless, the thermodynamic model describes the experimental data very well over a wide concentration range (see Fig. 2.18 in Chapter 2).

The 1:1 reaction stoichiometry can be represented by the following general equation ($\ln K_{EX} = 4.778$):

$$HA + \overline{TOA} \rightleftharpoons \overline{TOAHA} \qquad K_{EX} \tag{3.301}$$

The reaction stoichiometry has been discussed in Fig. 2.17, showing that the 1:1 complex prevails over the 1:3 one in the investigated concentration domain.

For the reaction kinetics the model of Danesi et al. (1978) with the following three molecular reaction steps was adapted:

$$H^+ + TOA_{ad} \rightleftharpoons TOAH^+_{ad} \qquad \text{qss: } K_1 \tag{3.302}$$

The index '*ad*' represents those species which are adsorbed at the interface. Therefore, the first reaction step depicts the protonation of an interfacially adsorbed *TOA*-molecule and is at a quasi-stationary equilibrium. The first step is instantaneous in contrast to the two following rate-determining reaction steps:

$$TOAH^+_{ad} + A^- \rightleftharpoons TOAHA_{ad} \qquad \text{rds: } k_2;\, k_{-2} \tag{3.303}$$

The second reaction represents the formation of the ion-pair molecule which is adsorbed at the interface. This molecule is replaced by a fresh *TOA*-molecule from the organic bulk phase in the third reaction step:

$$TOAHA_{ad} + \overline{TOA} \rightleftharpoons TOA_{ad} + \overline{TOAHA} \qquad \text{rds: } k_3;\, k_{-3} \tag{3.304}$$

The interfacially adsorbed anion exchanger species follow the Langmuir law:

$$[TOA_{ad}] + [TOAH^{+}{}_{ad}] + [TOAHA_{ad}] = \frac{\alpha_L\ [\overline{TOA}]}{\gamma_L + [\overline{TOA}]} \tag{3.305}$$

α_L and γ_L are the Langmuir constants which were determined by interfacial tension measurement (α_L=8.901×10^{-4} and γ_L=4.836×10^{-2} mol/kg). To guarantee the model consistency for the equilibrium state at infinite time, the activity, *a*, of the species is used. Combining the rate determining steps (equs. (3.301)–(3.304)) and equ. (3.305) yields the kinetic rate equation, R':

$$R' = \frac{(a_{H^+}\ a_{A^-}\ a_{\overline{TOA}} - 1/K_{EX}\ a_{\overline{TOAHA}})\ (\alpha_m\ a_{\overline{TOA}} / (\gamma_m + a_{\overline{TOA}}))}{C_1 + C_2\ a_{H^+} + C_3\ a_{H^+}\ a_{A^-} + C_4\ a_{H^+}\ a_{\overline{TOA}} + C_5\ a_{\overline{TOA}} + C_6\ a_{\overline{TOAHA}}} \tag{3.306}$$

The constants $C_1 - C_6$ are functions of the kinetic parameters K_1, k_2, k_{-2}, k_3 and k_{-3}:

$$C_1 = \frac{k_{-2}}{k_2 k_3 K_1},\ C_2 = \frac{k_{-2}}{k_2 k_3},\ C_3 = \frac{1}{k_3},$$
$$C_4 = \frac{1}{k_2},\ C_5 = \frac{1}{k_2 K_1},\ C_6 = \frac{1}{k_{-2} K_{EX}} \tag{3.307}$$

A detailed derivation of equ. (3.306) is given in appendix D. However, the only difference is that instead of equ. (D11), equ. (3.305) is used [Czapla (2000)]. The electrostatic extension of this kinetic model is as follows. The interfacial concentration of the aqueous species m_i^I which is caused by the induced electrical potential is derived according to the Boltzmann equation (equ. (3.188)) with the help of the exact solution of the Poisson–Boltzmann equation (equ. (3.194)). The results are depicted in Fig. 3.36. As can be seen the interfacial concentrations differ in magnitude from the bulk ones. Since the interfacial concentrations enter kinetic equations the decrease or increase of the mass transfer rate (see Fig. 3.34) is evident.

The model is able to calculate the enhancement and reduction of the transfer rate of the experimentally determined reactive mass transfer with kinetic parameters from the surfactant free system when taking the polarisation effects into account. Table E-9 (see appendix E) gives model parameters where k_{-3} can be derived from K_{EX} and k_3. Furthermore, the incorporation of the equilibrium constant in the kinetic model guarantees the transition of the kinetic model and its electrostatic extension to the equilibrium value for the boundary at infinite time. The comparison between the calculated and the experimental data shows that the kinetic model quantitatively predicts the enhancement and reduction of the mass transfer rate in the presence of cationic and anionic surfactants when extracting ionic solutes (see Fig. 3.37).

However, this analysis is only valid in a reaction dominated regime, which is usually at a low concentration range. At higher concentrations diffusion or even convection inside a droplet may dominate and aqueous side polarization efffects are thus neglegible.

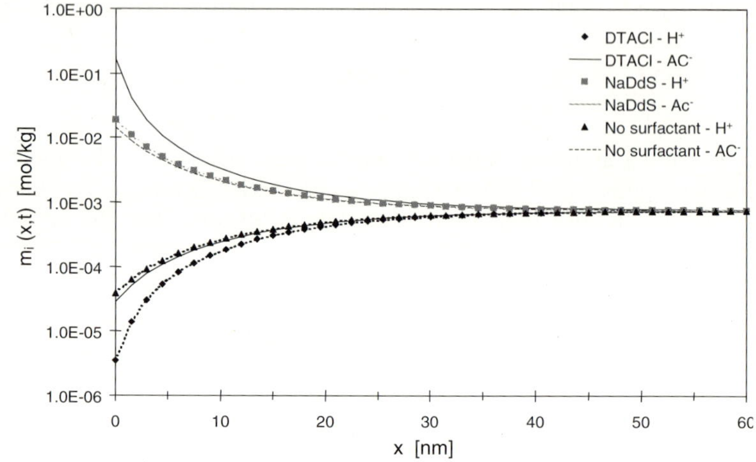

Fig. 3.36: Concentration profile in the polarisation layer

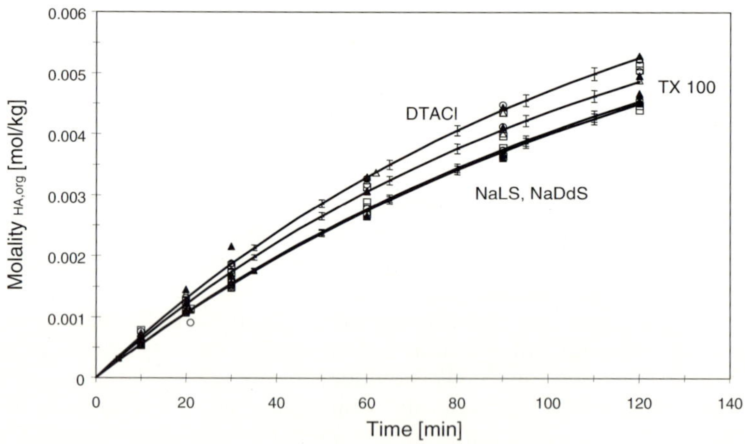

Fig. 3.37: Experimental and simulated concentration time curves (0.2 wt.% *HAc*, 10% *TOA* in toluene, interfacial coverage = 0.75 μmol/m^2 of DTACl, TX100, NaLS, and NaDds, T = 298 K)

A further modification in order to improve the consistency of kinetic models is with the LAP model [Roos & Bart 2000b)]. In technically relevant extraction systems the interfacial area is always totally covered by the amphiphilic ion exchanger molecules or other surfactants. For instance, the

TOA/H_2O/toluene system shows a constant interfacial coverage, γ, (constant slope according to the Gibbs equ. (3.176) as depicted in Fig. 3.38).

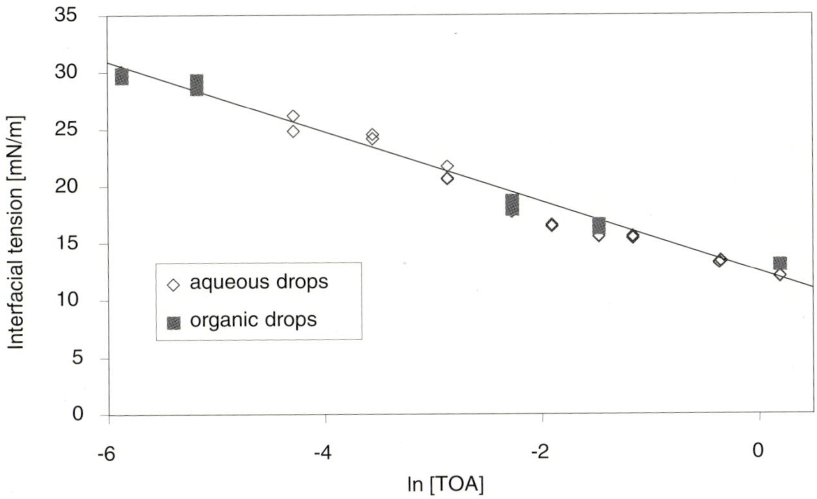

Fig. 3.38: Interfacial tension for the system *TOA*/H_2O/toluene at 298 K

Again, for the extraction of acetic acid the same three elementary steps (equs. (3.302)–(3.304)) give the overall reaction (equ. (3.301)). The protonation of the anion exchanger is instantaneous and the reactions with the solute are rate determining. The interfacially adsorbed species cover the available interfacial area in contrast to equ. (3.305) as to:

$$[TOA_{ad}] + m\,[TOAH^+_{ad}] + n\,[TOAHAc_{ad}] = \max . \tag{3.308}$$

Here m and n scale the different interfacial covered area of the compounds and are set as an approximation to unity. Due to this maximal coverage constraint there is a limited number of adsorption places which is the basis of the LAP-model. Since the protonation of *TOA* is instantaneous, the following reaction rate is valid:

$$R = \frac{[\overline{TOA}]\,[H^+]\,[Ac^-] - \dfrac{1}{K_{ex}}\,[\overline{TOAHAc}]}{denom}\ \max \tag{3.309}$$

with

$$denom = \frac{1}{k_2}\left\{\left[\frac{1}{K_{EQ1}}\overline{[TOA]}+\overline{[TOA]}[H^+]+\frac{1}{K_{EQ1}\,K_{EQ3}}\overline{[TOAHAc]}\right]\right\}$$
$$+\frac{1}{k_3}\left\{\frac{1}{K_{EQ1}\,K_{EQ2}}+\frac{1}{K_{EQ2}}[H^+]+[H^+][Ac^-]\right\} \quad (3.310)$$

A detailed derivation of equ. (3.309) is given in appendix D. The advantage is when there is a multicomponent extraction, the only change being that equ.(3.308) has to include the new co-extracted interfacially adsorbed species. In the corresponding rate equation (equ. (3.309)) nothing will change (see appendix D for the co-extraction of chloride and acetate), but the *denom* (equ. (3.310)) will be expanded. Thus a consistent modelling of multicomponent kinetics is possible and the parameters estimated from every individual single solute extraction system can be combined likewise to the multicomponent with only the mixing rule of equ. (3.308).

However, at higher acetic acid concentrations (as a special case with *HAc*) there is also the physical solubility to be taken into account. Acetic acid is mainly monomeric when dissociated in the aqueous phase and predominantly dimeric in the organic phase. A simple mass transfer model describes the physical extraction of acetate simultaneously to the chemical reaction (see equ. (3.311)) where the dimeric species is formed at the interface.

$$R_{dimer} = k_{dimer}\,(K_{EQ,dimer}\,[H^+]^2\,[Ac^-]^2 - \overline{[(HAc)_2]}) \quad (3.311)$$

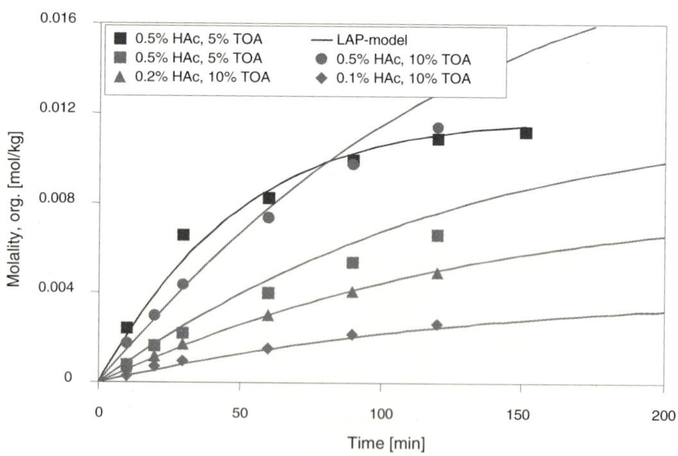

Fig. 3.39: Acetic acid extraction (• Roos (2000), remaining symbols Czapla (2000))

The quality of this model is depicted in Fig. 3.39 where data from different mass transfer cells [Roos (2000), Czapla (2000)] are included. The chloride extraction kinetics is depicted in Fig. 3.40 where a physical solubility similar to equ. (3.311) could be neglected.

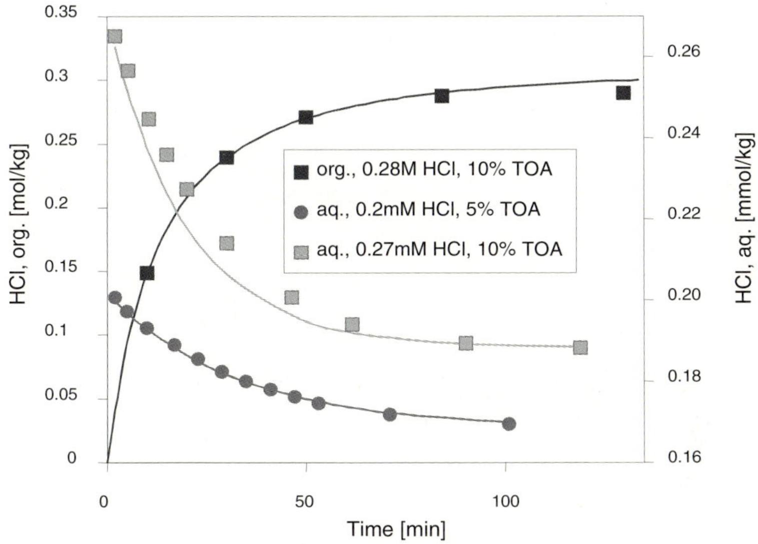

Fig. 3.40: *HCl* extraction kinetics at different initial conditions

A comparison of the measured data and the model predictions with parameters from the single solute system to the mixed system is given in Figs. 3.41a and b. As can be seen, the predictions are satisfied if we consider experimental errors with ± 5%.

a)

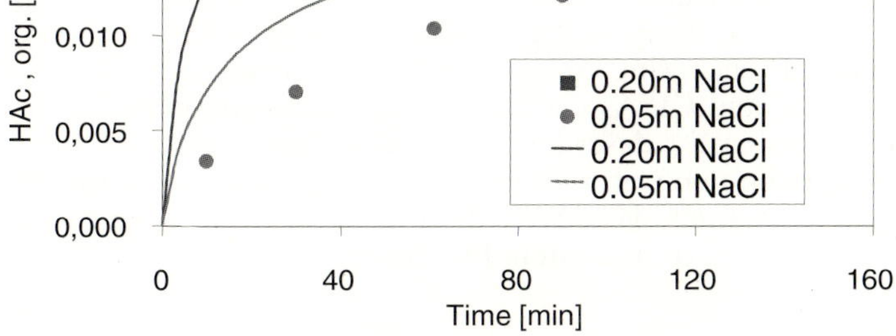

b)

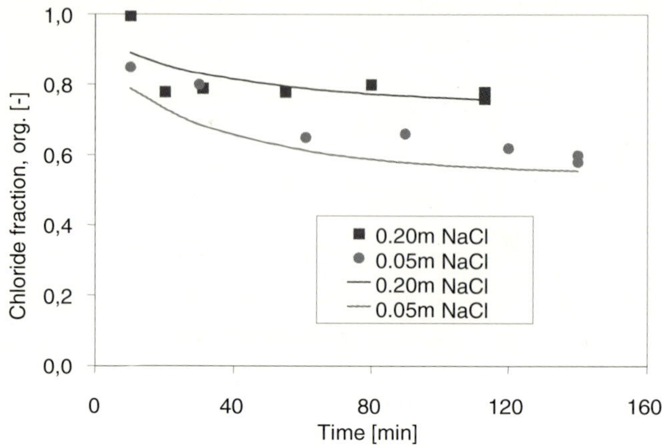

Fig. 3.41: LAP model and kinetic experiments with 0.2% *HAC*, 10 % *TOA* in toluene at different *NaCl* concentrations (298 K) *a* acetate extraction *b* chloride extraction

3.3.5 Macrokinetics of acid extraction with TOA

In a similar procedure as with *Zn*/*D2EHPA* the mass transfer in a Venturi-tube is modelled with the Maxwell–Stefan equations neglecting aqueous resistances. It is assumed that the solvent is insoluble in the aqueous phase ($N_n = 0$) and the system behaves in a pseudo-binary manner, that is the cross coefficients are zero. The matrix elements B then simplify to:

$$B_{i,i} = 1 / D_{i,n} \tag{3.312}$$

and

$$B_{ij} = 0 \tag{3.313}$$

The driving forces only consider differences in the chemical potential and the effective diffusion coefficient was modelled with the stagnant cap model (see equ. (3.167)) where $f_z = 0.14$. The initial and sampling effects during mass transfer in the Venturi-tube were eliminated when extrapolating to t=0 [Nitsch (1989)]. The measured and calculated mean concentration profiles within a single droplet could be fairly well correlated (see Fig. 3.42). The overall mass transfer coefficient according to equ. (3.296) is depicted in Fig. 3.43 (7 % relative error for d_r and k_y). As can be seen, the diffusion with the small rigid droplets cannot be easily described. Here with the neglect of the aqueous resistance, the assumption of a pseudo-binary diffusion in the organic phase results in these deviations. However, the transition from a rigid to a mobile surface is clearly shown (minimum in Fig. 3.43).

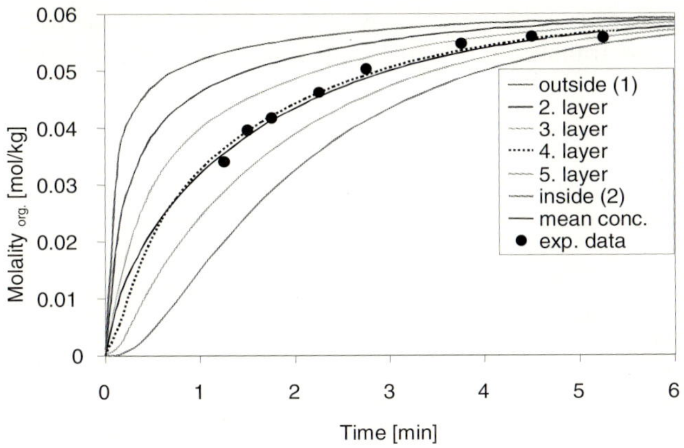

Fig. 3.42: Concentration profile in a droplet (d_p = 1.9 mm, 10% *TOA*, 1% *HAc*, 298 K)

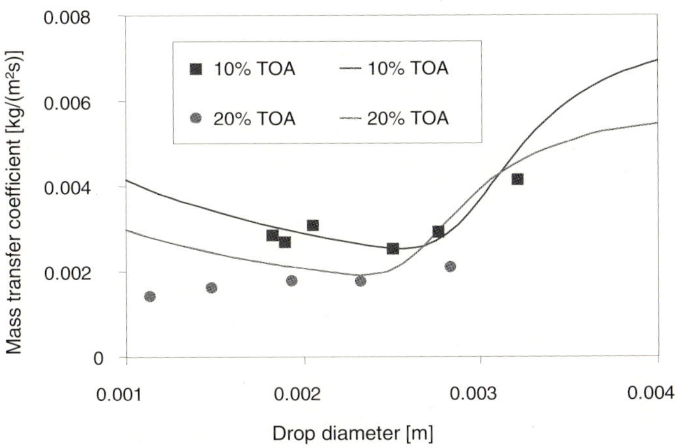

Fig. 3.43: Mass transfer coefficient (1% *HAc*, 298 K)

In a final step (compare Fig. 1.13) the single droplet mass transfer information was used to calculate the concentration profile of a 0.06 m diameter Kühni column (10 % open area, H_K = 2 m). Fig. 3.44 depicts the computed profiles, calculated with the mass transfer kinetics from the Venturi-tube experiments in comparison with a constant equilibrium stage efficiency of 39% (both phases are assumed to have plug flow in this small diameter laboratory column). A reasonable result can be achieved with the mass transfer model (about 15% rel. error) and systematic deviations are to be seen with the equilibrium stage model. The species efficiencies are depicted in Fig. 3.45 and are far from being constant along the column height. As can be seen, the apparatus design based on the equilibrium model, corrected with an efficiency factor, is risky especially when diffusional or reactive resistances dominate the mass transfer in reactive liquid–liquid extraction systems.

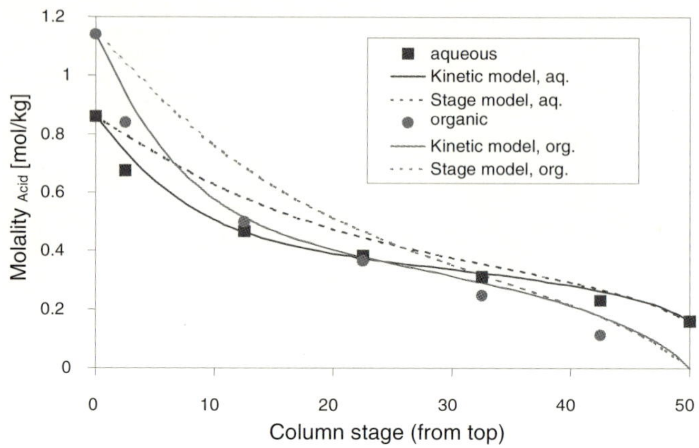

Fig. 3.44: Column concentration profiles

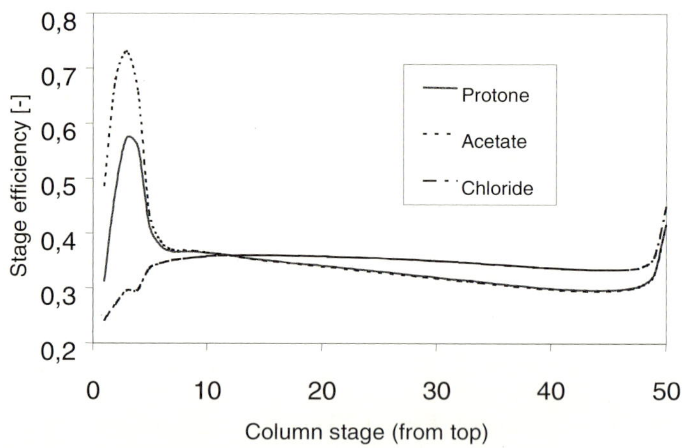

Fig. 3.45: Efficiency profile

4 Current Developments and Apparatus Techniques

4.1 Liquid–Liquid Contacting

The extraction process involves an intimate contacting of the two liquid phases in order to obtain reasonable mass transfer rates. Provision has to be made for separating the phases after completed contacting. Four major types of extractors are distinguished: *mixer–settler systems, column extractors, centrifugal extractors* and *non-dispersive membrane extractors.* Mixer–settlers are rather easy to design, flexible in residence time (an important issue with reactions), pH control in each extraction stage is individually possible, and even which phase dispersed may vary from one extraction unit to another. Due to this flexibility they dominate in the field of technical applications for reactive extraction. Centrifuges are very cost intensive (investment and maintenance) and are only used when there are phase disengagement or degradation problems (e.g. penicillin extraction from strong acidic media with residence times of a few seconds only [Ridgeway & Thorpe (1983)]). Here we will concentrate our discussion on non-dispersive membrane and counter-current column extractors. Two types of column extractors exist: non-agitated non-proprietary columns and agitated proprietary extractors. Agitating the liquid–liquid system breaks up droplets and increases the interfacial area to improve the mass transfer and column efficiency. Various forms of energy input are used:

- rotation of propeller, impeller, disc,
- pulsation, vibration, ultrasonic devices,
- centrifugal.

Some examples of mechanically agitated contactors are: Rotating Disk Contactor (RDC), asymmetric RDC, Kühni, Oldshue–Rushton and York–Scheibel columns as can be seen in Fig. 4.1.

There are three types of non-proprietary non-agitated types of extraction columns (see Fig. 4.2). The spray columns are the simplest type of extractors containing only distributors for the feed (often through perforated pipes). This makes them very inexpensive but they are limited in use due to the large axial mixing in the column and the fact that the phases are not coalesced and redistributed.

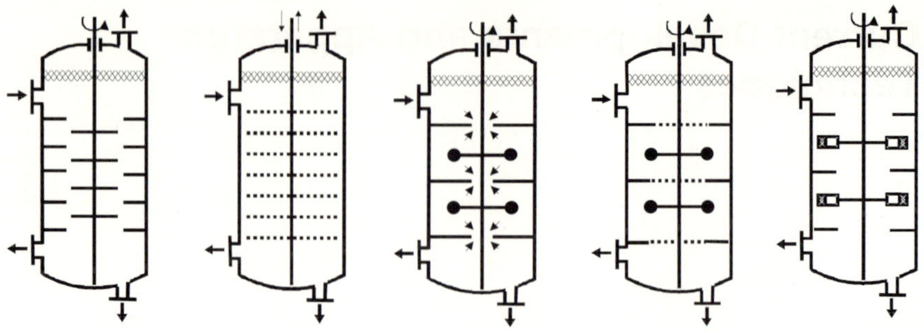

Fig. 4.1: Agitated extractors (left to right: RDC, Karr, Oldshue–Rushton, Kühni, Scheibel extraction columns)

This often results in low efficiencies, which are comparable to one or two theoretical equilibrium stages. Packed columns are much more efficient since the packing reduces backmixing and enhances drop reformation. Structured packings, in contrast to unstructured packings, reduce the cross-sectional area for liquid flow, giving rise to smaller diameters. The packing types that can be used are the same as those for normal distillation operations (e.g. rings, saddles or slightly modified structured packings of corrugated metal sheets). Sieve tray columns resemble the distillation column design with the exception of the downcomer design and that there is no weir. Also, in distillation the vapour always moves up the column, while in an extraction process either phase, the light or the heavy one, can be dispersed. This means that there are also two sieve tray designs. The hydraulic design of a sieve tray is given in appendix C.

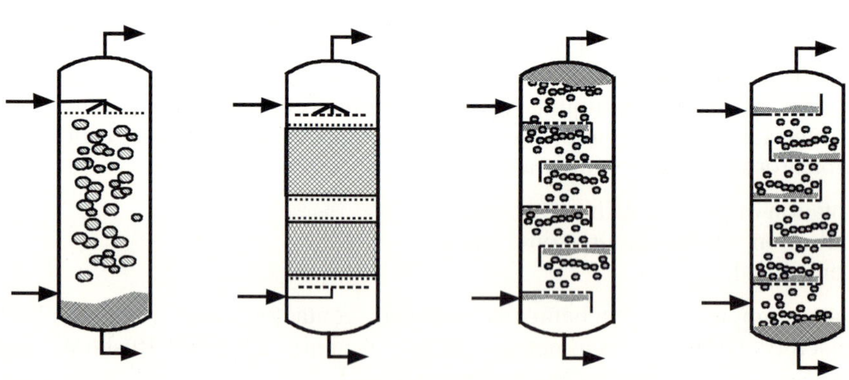

Fig. 4.2: Non-agitated extractors (left to right: spray, packed, sieve tray (light), sieve tray (heavy) column)

The selection of a specific contactor is complicated due to the large number of types available and the number of design parameters. Mechanically agitated devices show some efficiency advantages but whether these advantages come at the expense of higher cost must be evaluated by designing the specific system at hand.

The equipment size – column diameter and height – depends on which phase is dispersed (through the physical properties of the phases) and on the solvent *S* and feed flow rates *F*. Increasing the *S/F* ratio decreases the column height but increases its diameter. It also increases the cost of solvent recovery due to the lower solute concentrations in the extract. The column height depends on the separation requirements (extract and raffinate purities) and especially on the interphase mass transfer. The mass transfer is also a complicated function of the column hydrodynamics (drop size, rise velocity, mixing, etc.).

Upon deciding which phase to disperse we should also consider the wetting characteristics of both phases. The continuous phase should preferably wet the internals otherwise the dispersed phase will coalesce which lowers the interfacial area. Furthermore the interfacial area will be higher if the larger flow is dispersed and the mass transfer rates tend to be higher going from the dispersed to the continuous phase. Since the hold-up of the dispersed phase is lower than that of the continuous phase, a cost-intensive or dangerous phase is preferably dispersed. However, often plant praxis (settling problems, crud[1] formation and emulsification) dictates which phase is to be dispersed.

The hydrodynamics modelling is quite simple with mixer settlers. Usually the mixer can be considered as ideally stirred and since at each stage temperature, pH value, phase ratio (due to recycled flows from the settler), and organic or aqueous phase dispersion can be set individually, it is the predominant apparatus in hydrometallurgy. In the chemical and oil industry columns prevail and (pulsed) sieve trays exhibit an almost plug flow behaviour. Stirred columns achieve higher efficiency, but deviations from plug flow are the rule. The simplest way to account for these deviations is the dispersion model where one parameter, the axial dispersion coefficient D_{ax}, accounts for the non-ideal flow behaviour of each phase:

$$\frac{\partial c(t,z)}{\partial t} = -w\frac{\partial c(t,z)}{\partial z} + D_{ax}\frac{\partial^2 c(t,z)}{\partial z^2} - n(t,z) \tag{4.1}$$

The first term reads for non-steady-state, the second for convective plug flow, the third for non-ideal flow and the last for mass transfer. However, a problem arises applying this equation to the organic phase which has to be then considered quasi-homogeneous. This implies the same radial concentrations and one mean Sauter diameter droplet representing the droplet swarms moving through the column. In order to overcome this problem a more realistic model was developed, the forward mixing model [Rod (1966), Olney (1964)], which takes the different rising velocities of the different droplet sizes into account. Nowadays it is the basis for the so-called droplet population balance models developed with crystallisation [Hulburt & Katz (1964)] and first applied to liquid dispersions by Valentas and Amundson (1966), with regards to the kinetic processes like droplet rising, backmixing, break-up and coalescence [Ortner et al. (1995)]. This is discussed for extraction columns with the program POPMOD (http://www.uni-kl.de/LS-Bart/DAE).

[1] crud: Chalky River unidentified deposit [Ritcey (1998)]

Let $P(t,z,d)\mathrm{d}d$ be the volume fraction of drops of diameter between $d \pm \Delta d/2$ at column height z at time t. $P(t,z,d)$ may also be interpreted as the probability density to find a droplet of diameter $d \pm \Delta d/2$ located at time t at the column level z. The local hold-up of the dispersed phase can be calculated from $P(t,z,d)$ as follows:

$$\phi(t,z) = \int_0^{d_{max}} P(t,z,d)\,\mathrm{d}d \tag{4.2}$$

A volume or mass balance for droplets of diameter $d \pm \Delta d/2$ leads to the following population balance equation (without mass transfer):

$$\frac{\partial P(t,z,d)}{\partial t} = w_d \frac{\partial P(t,z,d)}{\partial z} + D_{d,ax} \frac{\partial^2 P(t,z,d)}{\partial z^2} + S_B(t,z,d,P) + S_C(t,z,d,P) \tag{4.3}$$

The first term on the right hand side of equation (4.3) represents the transport of droplets due to drop diameter dependent rising velocity w_d, the second part describes the axial dispersion of droplets. The source terms S_B and S_C take the birth and death of droplets with size $d \pm \Delta d/2$ due to breakage and coalescence into account. One has to consider at least ten droplet classes to get reasonable results [Kronberger et al. (1995)].

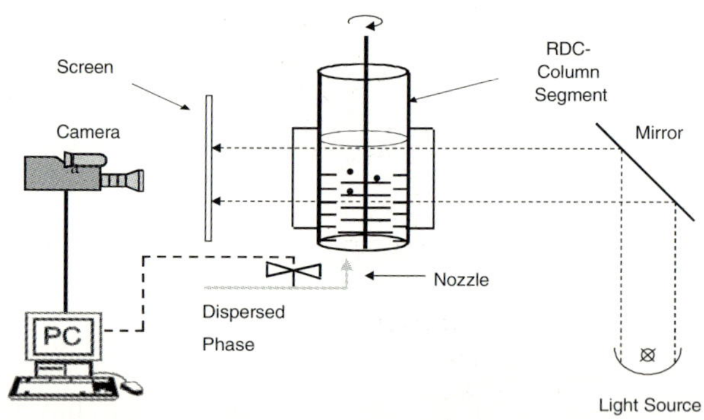

Fig. 4.3: Breakage experiments in a laboratory unit

The model seems rather sophisticated with the number of droplet specific parameters involved. Fortunately, at industrial relevant stirring speeds and (high) throughputs, D_{ax} is independent of the droplet size and approaches the value for the continuous phase [Modes (2000)] (the latter is usually easier to measure and well documented). Commonly in practice are found surfactant containing systems, which reduce surface tension and generate small rigid droplets with low tendency to coalescence. Thus the remaining main effect is the break-up of droplets which can be correlated or experimentally determined in a small cell which is designed geometrically similar to the column. Thus minimal aqueous and organic phase quantities are necessary to measure all design parameters (see Fig. 4.3).

Drops break-up into several smaller drops due to the turbulent structure of the flow pattern and the column internals. This situation has been extensively studied in stirred tanks (e.g. Coulaloglou & Tavlarides (1976)). When considering drops of diameter $d \pm \Delta d/2$, one will see loss as well as gain of such drops due to breakage. The loss due to breakage is given by:

$$S_B^-(t,z,d,P)\Delta d = g(t,z,d)\; P(t,z,d)\Delta d \tag{4.4}$$

The production of drops of diameter between $d \pm \Delta d/2$ from breakage of larger drops with d_0 is then given:

$$S_B^+(t,z,d,P)\Delta d = \Delta d \int_0^{d_{max}} \beta(d_0,d)\; g(t,z,d)\; P(t,z,d)\mathrm{d}d_0 \tag{4.5}$$

Droplet breakage is denoted by the parameters $\beta(d_0,d)$ and $g(d)$, where $\beta(d_0,d)\Delta d$ is the probability of production of a droplet with diameter $d \pm \Delta d/2$ due to the breakage of a droplet with diameter d_0 and $g(d)$ is the relative rate of drop breakage or the breakage frequency. According to Modes (2000) the relative drop break probability based on a modification of a correlation of Cauwenberg (1995) in a RDC column ($D_k = 0.15$m) is:

$$\frac{p}{1-p} = 0.000604\, We_{m,Ca}^{1.595} = 0.000604 \left(\frac{\rho_C^{0.8} \eta_C^{0.2} d\, D_R^{1.8} (N^{1.8} - N_{crit}^{1.8})}{\sigma} \right)^{1.595} \tag{4.6}$$

with the critical rotor speed

$$N_{crit} = 0.738 \left(\frac{\rho_C D_R^3}{\sigma} \right)^{-0.5} \left(\frac{d}{D_R} \right)^{-0.986} \tag{4.7}$$

Here ρ, η, σ are the density, the viscosity and the interfacical tension, respectively, and N and D_R are the rotor speed and the rotor diameter (in SI-units). The latter equation can be rearranged to obtain the critical droplet diameter where breakage starts at the critical rotor speed:

$$d_{crit} = 0.735\, D_R\, We_R^{-0.507} \tag{4.8}$$

For the concentration of the hold-up profile with the program POPMOD [Kronberger et al. (1995)] one needs the following correlations which are specific for the geometry of a RDC column. The axial rising velocity is:

$$w_d(t,z,d) = k_f(d)\, w_T(d)(1-\phi)^m - \frac{Q_c}{A\,(1-\phi)} \tag{4.9}$$

where the parameter m is set to unity and the last term denotes the continuous phase flow (Q_c) with A the free column area and the hold-up, ϕ. The terminal velocity, w_T, is [Vignes (1965)]:

$$w_T(d)=\frac{d}{4.2}\left(\frac{g\Delta\rho}{\rho_c}\right)^{2/3}\left(\frac{\rho_c}{\eta_c}\right)^{1/3}\left(1-\frac{Eö}{6}\right) \tag{4.10}$$

and the friction factor k_f is a function of the diameter of the stator, D_S, of the rotor, D_R, and of the stirring speed, N [Modes (2000)]:

$$k_f=1-1.037\left(N^3D_R^5\right)^{0.12}-0.62\left(\frac{d}{D_S-D_R}\right)^{0.44} \tag{4.11}$$

The axial dispersion coefficient of the continuous phase, $D_{ax,c}$, for a characteristic column length, H_c, is after Kumar & Hartland (1994):

$$\frac{1}{Bo(c)}=\frac{D_{ax,c}}{w_cH_c}=0.42+0.29\left(\frac{w_d}{w_c}\right)+\left[c_1\left(\frac{N\,D_R}{w_c}\right)+\frac{13.38}{3.18+\left(\frac{N\,D_R}{w_c}\right)}\right] \tag{4.12}$$

$$\left(\frac{w_cD_R\rho_c}{\eta_c}\right)^{-0.08}\left(\frac{D_K}{D_R}\right)^{0.16}\left(\frac{D_K}{H_c}\right)^{0.1}\left(\frac{D_S}{D_K}\right)^{2}$$

and for the dispersed phase [Modes (2000)]:

$$\frac{1}{Bo(d)}=\frac{D_{ax,d}}{w_dH_c}=2.76\quad 10^{-3}+1.65\left(\frac{N\,D_R}{w_d}\right)^{3.3} \tag{4.13}$$

where H_c is the compartment height and D_K the diameter of the column.

The coalescence term, S_c, in equ. (4.3) is about zero in coalescence hindered systems (e.g. toluene/water). The breakage frequency $g(d)$ of a single droplet can be calculated by dividing this probability by the mean residence time of a droplet in one column compartment with height H_c. This leads to the following expression for $g(d)$:

$$g(d)=p(d)\frac{w_d(d)}{H_c} \tag{4.14}$$

The probability density, $p(d)$, is according to equs. (4.6) and (4.8). Bahmanyar and Slater (1991) suggest the use of a beta distribution function in order to relate the daughter droplet distribution produced from the breakage of a mother droplet:

$$\beta(d_0,d)=3\,x_m\,(x_m-1)\left[1-\left(\frac{d^3}{d_0^3}\right)\right]^{(x_m-2)}\frac{d^5}{d_0^6} \tag{4.15}$$

where after Modes (2000) the medium number of daughter droplets is:

$$x_m = 2 + 0.17\left[\left(\frac{d_M}{d_{crit}}\right) - 1\right]^{1.83} \tag{4.16}$$

On the basis of this model a comparison of experimental and simulated droplet size density distributions in a RDC column for the system toluene/water (D_K = 0.15 m, D_S = 0.105 m, D_R = 0.09 m, H_c = 0.03 m, active column height = 2 m) is given in Fig. 4.4. The corresponding hold-up profile is depicted in Fig. 4.5. As can be seen, due to the droplet interactions there is a changing mean Sauter diameter, d_s. The large inlet droplets decrease continuously. The local hold-up changes slightly and the interfacial area increases from the inlet with about 130 m^2/m^3 to about 260 m^2/m^3 according to the relation:

$$a(z) = \frac{6\,\phi(z)}{d_S(z)} \tag{4.17}$$

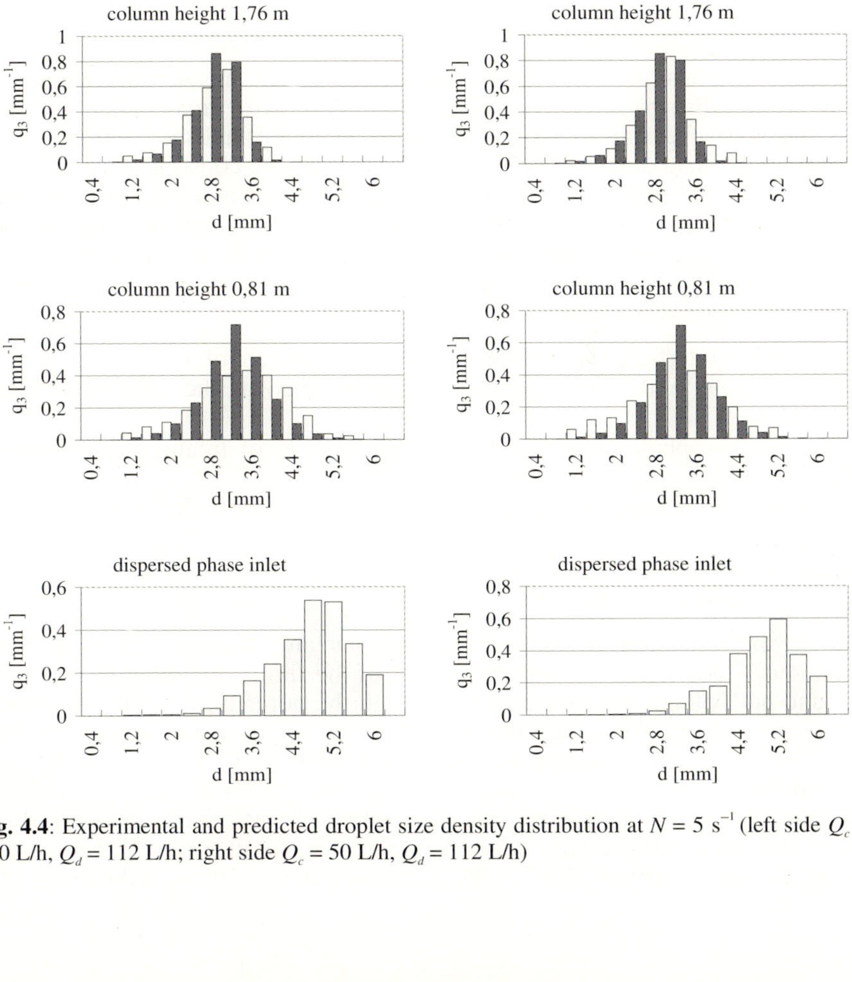

Fig. 4.4: Experimental and predicted droplet size density distribution at $N = 5\ s^{-1}$ (left side Q_c = 100 L/h, Q_d = 112 L/h; right side Q_c = 50 L/h, Q_d = 112 L/h)

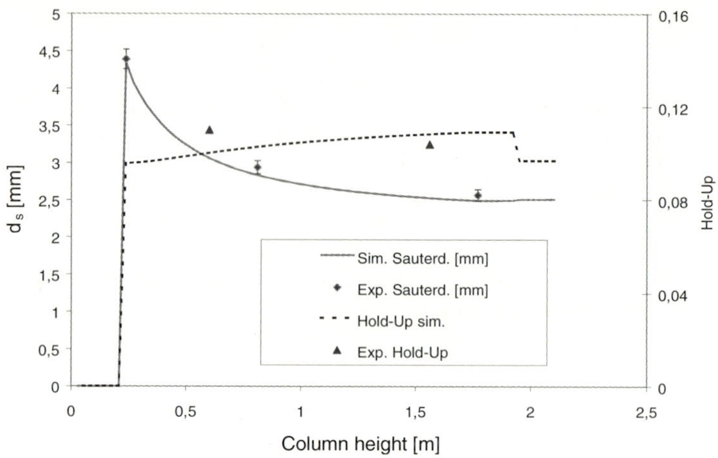

Fig. 4.5: Hold-up and mean droplet diameter profile at $N = 5\ s^{-1}$, $Q_c = 100$ L/h and $Q_d = 112$ L/h (toluene/water)

The method lets one calculate the local hold-up, the mean Sauter diameter or the droplet distribution and predicts the mass transfer area available in the apparatus. However, since this method is not widely established and tested in industry, pilot plant data are commonly used for design, construction and start-up of an industrial plant which to some extent is a costly and lengthy exercise.

An alternative to conventionally performed counter-current liquid–liquid extraction is to use *hollow fibre modules*. This membrane extraction process is also known as pertraction and combines the advantages of a membrane process with a solvent extraction process i.e.:

- a very high interfacial area allows a high recovery rate of a solute,
- high aqueous and organic phase velocities even at the case of similar densities,
- allowing any phase ratio and being thus capable of treating very diluted feeds,
- minimal solvent loss and virtually no entrainment due to the non-dispersive contact,
- reduction in equipment volume and easy scale-up,
- easy operability and control

In the hollow fibre contactors the aqueous and organic solutions flow continuously, one phase through the lumen of the fibres, the other at the shell side, and both phases get into contact through the pores of the fibre. A proper adjustment and control of the differential static backpressure will leave the interfacial contacting area in the pores and will avoid bleeding and thus entrainment.

Extensive reviews on hydrodynamics, wetting characteristics, backpressure design and mass transfer correlations are given in the literature [D'Elia et al. (1986), Prasad & Sirkar (1987, 1992), Seibert et al. (1993), Schöner et al.

(1998), Costello et al. (1993), Wickramasinghe et al. (1992), Yang & Cussler (1986), Reed et al. (1995), Seibert & Fair (1997), Wang & Cussler (1993), Gabelman & Hwang (1999), Yoshizuka et al. (1986, 1995)].

The technique can be applied to similiar problems as conventional reactive solvent extraction, e.g. metal separation and recovery [Kim (1984), Kathios et al. (1994), Yang et al. (1996), Alonso et al. (1994), Soler et al. (1996), Matsumoto et al. (1990)], effluent treatment [Daiminger et al. (1996), de Haan et al. (1989), Yun et al. (1989, 1993), Reed et al. (1994), Sato et al. (1990), Basu et al. (1990), Cooney & Jin (1985)] and extraction of acids, pharmaceuticals and bio-products [Basu & Sirkar (1991, 1992), Dahuron & Cussler (1988), Prasad & Sirkar (1988, 1989), Coelhoso et al. (1997), Matsumura & Märkl (1986), Matsumura (1991), Naser & Fournier (1988), Lopez & Matson (1997), Ding et al. (1992), Escalante et al. (1996), Fish et al. (1992), Vatai & Tekic (1991), Shukla et al. (1989].

The commercially available hollow fibre liquid–liquid membrane contactors [Sengupta et al. (1994)] are operated in co- or cross-current flow. The latter operation method is achieved with baffels (see Fig. 4.6), which then enhances mass transfer [Wickramasinghe et al. (1992), Schöner et al. (1998)]. The microporous PP fibres are potted in a solvent resistant epoxy or polyethylene tubesheet and the shell casing is PP, PVDF or stainless steel. They allow a throughput from 0.2 to 0.6 m^3/h and are available from, e.g. Celgard [Sirkar (1997)] or TNO [Feron et al. (1993)]. In 1998, TNO installed the first full scale 15 m^3/h pertraction unit. In general, the cost situation is advantageous compared to air stripping or activated carbon treatment as, f. e., for a 10 m^3/h feed stream with 10 mg/L trichloroethene the total costs per m^3 for an effluent with 1000 µg/L is 3.1 • and with 1µg/L is 5.6 • [Klaasen & Jansen (1998)]. Likewise the TNO fibre modules the Liqui-Cel Contactors from Celgard are mainly used for aqueous gas–liquid applications [Gabelman & Hwang 1999] as there is no attack from organic solvents (swelling ...). As can be seen, there is a large amount of current research on non-dispersive extraction, but as yet industrial application is not widely spread.

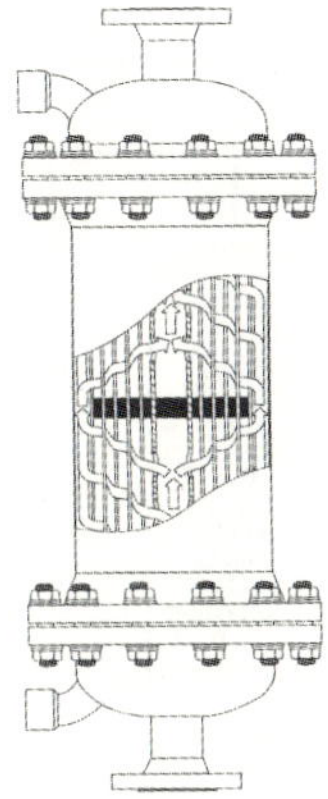

Fig. 4.6: Liqui-Cel cross flow membrane module

4.2 Polymeric Extraction Systems

For the last several decades, separation scientists have been employing the use of polymers to effect the extraction of a wide variety of target molecules. The polymers are either soluble

- in the organic phase,
- in the aqueous phase, or
- not soluble at all.

In the first case we denote non-ionic surfactants, which will form reversed micelles [Abou-Nemeh & Bart (1998)] and are used with *liquid surfactant membrane processes* [Marr et al. (1981)]. In the second case we have ionic and non-ionic surfactants which are used in *aqueous biphasic systems*, which is a gentle extraction medium for biomolecules being extracted without denaturation [Schick & Fowkes (1966–1999)]. The last case denotes the use of polymeric porous resins which can be impregnated with liquid ion exchangers and can be used in *extraction chromatography* [Warshawsky (1979)].

The *liquid surfactant membrane process* is an alternative to solvent extraction and ion exchange at highly diluted solute concentrations (usually less than 1 g/L) and is able to produce raffinates in the ppm level within one stage. The surfactants, like SPAN 80 (sorbitanmonooleate) or ECA 4360 (n-oligo(ethylenimino)succinimido-polyisobutylene) are dissolved in the organic phase II, which is recycled and a make-up in a process circuit is therefore at a minimum. The aqueous stripping phase I is emulsified in the organic phase. In an emulsion liquid membrane process (see Fig. 4.7) this emulsion is dispersed in the aqueous feed solution III and after mass transfer it is split in an electric field and the organic phase is reused. The mass transfer or pertraction is performed in mixer–settlers or in extraction columns as long as there is a proper density difference for dispersed emulsion phase buoyancy and phase separation in a settler.

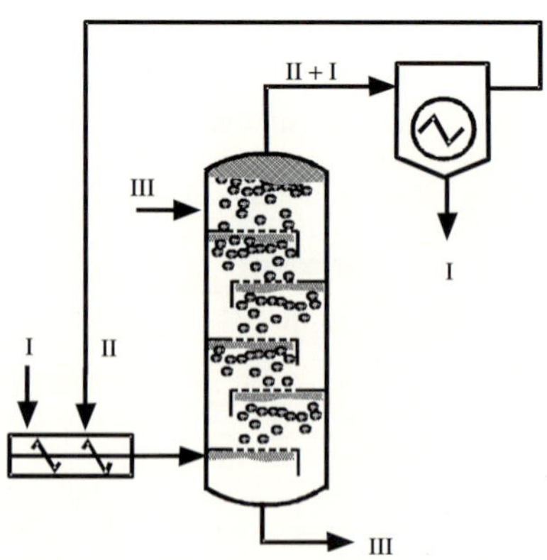

Fig. 4.7: Liquid membrane circuit (static mixer, permeation column and electrostatic emulsion splitter)

A detailed discussion of industrial applications is given by Marr & Draxler (1992). A special advantage of this technique is the use of the facilitated transport [Cussler (1971), Lorbach et al. (1986)] which greatly helps with very diluted feed streams. As can be seen in Fig. 4.8, the facilitated transport concentrates the diluted solute of the feed solution into the stripping solution according to the phase ratio of strip solution to feed solution.

$$Zn^{2+} + \overline{(D2EHPA)_2} \Leftrightarrow \overline{Zn(D2EHPA)_2} + 2H^+$$

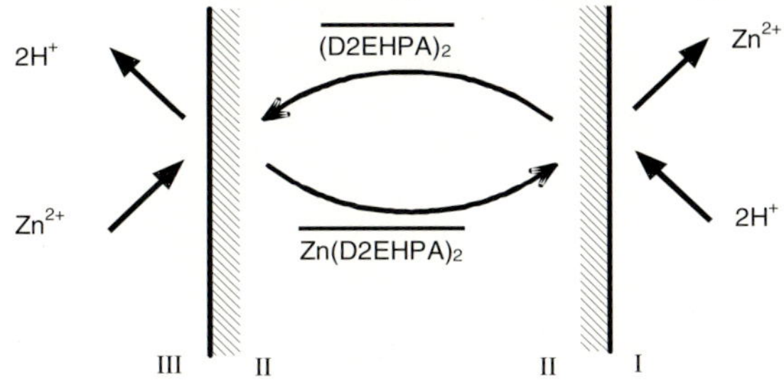

$$\text{Equilibrium:} \quad \frac{Zn_I^{2+}}{Zn_{III}^{2+}} = \left(\frac{H_I^+}{H_{III}^+}\right)^2$$

Fig. 4.8: Facilitated transport for zinc with *D2EHPA*

The physical basis of the facilitated transport is a coupled transport of the solute and a facilitating species. Both must be related or constrained, as for example by the overall extraction reaction. The depicted reaction is valid at the emulsion membrane globule outer interface (III/II), where the solute is extracted. The same relation holds at the inner membrane surface (II/I), where stripping is achieved with an excess of protons. This high driving force of the (facilitating) protons from the stripping phase to the outer aqueous feed phase facilitates the transport of the solute diffusing with the cation exchanger (or carrier) in the opposite direction to the protons. The maximal zinc enrichment is bound to the square of the proton driving force between phase I and III. A survey of the different mechanisms and theory is given by Way and Noble (1992) and Ho and Li (1992).

However, a part of this driving force is diminished, since inverted micelles (see Fig. 4.9) transport water from the outer to the inner aqueous phase [Bart et al. (1995)]. When there is a diluted feed solution at low ionic strength and an acidic stripping solution of high ionic strength (e. g. 2.5 M H_2SO_4), a large osmotic difference occurs which causes a swelling and, within minutes, a 100 % increase of the inner stripping phase [Bart et al. (1988, 1993)]. In order to avoid these inverted micelles, the extraction process can also be performed in solid membrane modules, which is a supported liquid membrane process where surfactant is not required [Danesi et al. (1987)]. The organic phase is adsorbed into the pores of a solid membrane. The stripping phase is in the lumen of the fibres and the feed phase on the outer shell. However, with an improper pressure balance there will be bleeding and a loss of the organic phase from the pores of the membrane which will result in a shut-down of the operation. As can be summarised, the technique offers major advantages with respect to diluted feed solutions due to the facilitated transport mechanism but is not promising under all circumstances (e.g. large differences in ionic strength between feed and stripping phase).

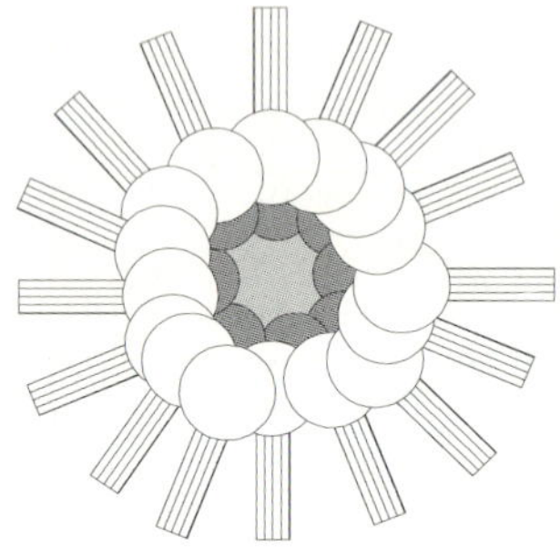

Fig. 4.9: Inverted micelle (core contains aqueous phase)

In comparison with the liquid surfactant process, neutral and ionic surfactants are used in *aqueous biphasic systems*. In contrast to emulsion liquid membranes they are added to the aqueous feed solution, which always causes losses with the raffinate. According to the kind of surfactant used [Huddleston et al. (1991)] we find cloud-point extraction [Hinze & Pramauro (1993)], micellar extraction [Scamehorn & Harwell (1988)], thermoseparating polymers [Johansson et al. (1995)], which are all special cases of aqueous biphasic systems [Huddleston et al. (1999)], and further on we have microemulsions [Leung et al. (1988)] and inverted micelles [Hurter & Hatton (1992)]. Microemulsions are thermodynamically stable and may exhibit a density between a pure aqueous and a pure organic phase. They are classified according to the Winsor types. Their existence is a function of the temperature and concentrations involved [Kahleweit et al. (1990), Overbeck et al. (1987)] and there is always a coexistence of an organic and aqueous bulk.

Reversed micelles do not only create the osmotic flux in emulsion liquid membranes but can also solubilise proteins and other solutes (Nitsch et al. (1996) in their aqueous environment. Since they exist in a conventional organic extraction phase, the apparative technique is quite similar to solvent extraction, where a change in temperature destroys the micelles and affects reextraction [Goklen & Hatton (1986)].

All the aqueous two phase systems depend on poly(oxyalkylenes) like poly(ethylene)glycol, *PEG* and similar polymers, to be effective. The shift between extraction and reextraction is achieved either by change in temperature or ionic strength. Classic aqueous biphasic systems are formed of either mixtures of polymer 1 (e.g. *PEG*) and polymer 2 (e.g. dextran) or polymer 1 and salt (e.g. K_3PO_4) [Walter & Johansson (1994), Rogers & Eiteman (1995)] above their critical concentrations. A split in two aqueous phases with decreasing temperature is related to the upper critical solution temperature (UCST, see Fig. 1.6) with *PEG* polymer systems

[Ananthapadmanaban & Goddard (1987), Van Oss et al. (1989)] and with increasing temperature (LCST, see Fig. 1.6) with *PEG*/salt systems [de Belval et al. (1998)]. While the applications to biomolecule partitioning are particularly well known [Großmann et al. (1995)], metal ions can also be separated [Rogers et al. (1995)]. The distribution of charged species depends on the pH value and the distribution of uncharged species follows the order of partitioning in 1-octanol /water systems [Eiteman & Gainer (1992)].

Cloud point extraction systems and thermoseparating polymers, in contrast to aqueous biphasic systems, are always known as micellar. An increase in temperature or change in salt concentration will lower the solubility of the non-ionic surfactant and at its lower consolute point (cloud point) two phases will appear: a micelle free-solution of surfactant close to its c.m.c. and a less dense upper phase depleted in surfactants. Thermoseparating polymers (*PEO–PPO* copolymers) form micelles at lower concentrations and therefore have lower c.m.c.'s as cloud point extraction systems (alkyl ethoxylates, poly (oxyethylene) ethers), which reduce the amount of polymer required to achieve phase separation. The use of cloud-point extraction for nickel with Triton X-100 as surfactant and 1-(2-thiazolylazo)-2-naphthol as extractant was first reported by Ishii et al. (1977). The solvent-like properties of polymeric solubilizing systems may vary considerably. On a solvatochromic scale [Doug & Winnik (1984)] Triton X-114 micelles are equivalent to the solvent 1,2 dichloroethane at 5° C and an increase of temperature to 25°C decreases the polarity equivalent to the solvent 1,5 pentanediol. In order to reduce losses the process may be performed in hollow fiber membranes [Lebens & Keurentjes (1996), Hurter & Hatton (1992)]. The membrane separates the contaminated feed stream from the receiving polymer solution and only the solute but not the polymer can pass the membrane and be pertracted. Alternatively, such polymers can be covalently bound to a supporting chromatographic stationary phase and thus losses of the polymer in operation can be avoided [Rogers et al. (1996)].

However, a more flexible and more specific separation can be achieved with *extraction chromatography*, when impregnating liquid ion-exchangers on polymeric water insoluble resins. The recovery of substances from highly diluted aqueous phases with liquid–liquid extraction is difficult because the amount extracted in one equilibrium stage is usually rather small. The use of solvent-impregnated resins combines the advantages of reactive liquid–liquid extraction and ion exchange in fixed beds:

- the liquid ion exchanger is adsorbed undiluted, which results in a higher capacity,
- the liquid ion exchangers are cheap and easily available, which generates wide flexibility in solute extraction,
- even mixtures of liquid ion exchangers might be used,
- the operation of a fixed bed is simple and the feed might be discontinuous,
- the bleeding loss of the ion exchanger is always less than entrainment losses in liquid–liquid systems.

However, there is a necessity to install two columns, one for extraction and one for stripping/regeneration in order to achieve a continuous operation. In general, the mass transfer is slower compared to a liquid–liquid extraction process, since the solute has to diffuse into the porous support material. Inert cheap macroporous materials are available (e.g. DELOXAN AP II®, Degussa Hanau or XAD®, Rohm & Haas) and the impregnation [Juang & Chang (1995)] is easily performed in situ in the fixed bed column [Traving (2000)]. The support materials are either gel-analogous (e.g. Amberlite XE-305®) or macro-porous inert (e.g. XAD2®, XAD4®, XAD16®) or functionalised resins (e.g. DELOXAN AP II®). In the first alternative the ion exchanger is only on the surface, which is not very efficient. The porous materials start to adsorb the ion exchanger in the smallest pores and also in the bigger ones with higher loadings [Strikovsky et al. (1996)]. An additional functional group may interact with the ion exchanger and thus reduce bleeding [Warshawsky & Patchronik (1978)] or may be actively extracting a solute [Warshawsky & Berkowitz (1979)]. Fig. 4.10 gives a REM-picture of the surface of an inert grain. The left side shows an unimpregnated, the right side an impregnated grain. The liquid ion exchanger that can be seen at the surface is swept away in a washing step and in all further operations the bleeding is almost negligible. Fig. 4.11 shows, that after the first cycle the losses will reduce perceptibly and are always beyond the physical solubility limit of *TOA* in water (7 ppm) due to the adhesive forces.

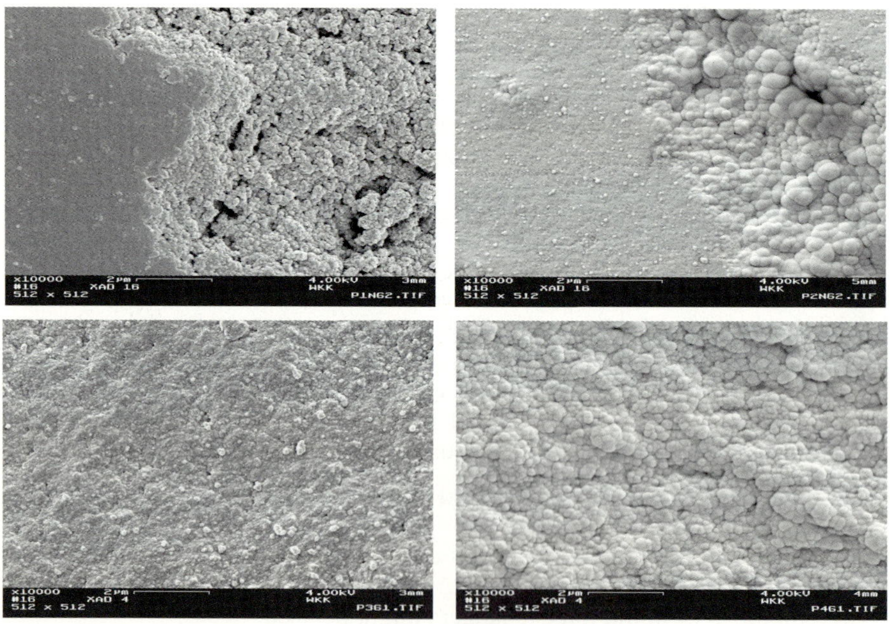

Fig. 4.10: REM picture of fresh and impregnated amberlite resin with (right) and without (left) ion exchanger (top: outer surface XAD16, bottom: cut particle XAD4)

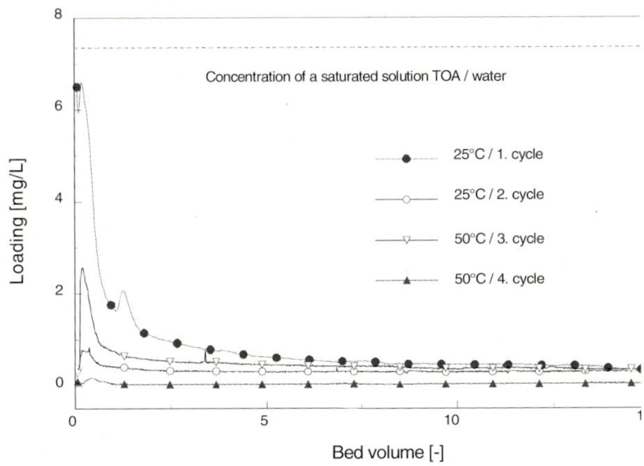

Fig. 4.11: Bleeding (XAD 16, initial loading of 520g *TOA*/ kg resin)

Alternatively the ion exchanger can be admixed during the synthesis of the resin [Kroebel & Meyer (1971)] and such resins are known as Levextrel resins (Bayer Leverkusen). This procedure affects the resulting porous structure and gives different results when different ion exchangers are used.

The theoretical description of the behaviour of solvent-impregnated resins can be subdivided into three major parts: the equilibrium situation, the mass transfer within one grain and the behaviour of the fixed bed adsorber. Most of the literature is related to equilibria and is descriptive and not predictive (see table 4.1).

Table 4.1: Models for solvent impregnated resins

Literature	Resin	Ion exchanger	Ion	Equilibrium	Kinetics
Akita (1990)	XAD 2	*TOA*	*Zn, Cu*	slope analysis	qualitative
Akita (1993)	XAD 2, XAD 4	*TOA*	*Pd*	Langmuir isotherm	
Cortina (1996)	Levextrel	*DTMPPA*[2]	*Zn,Cu,Cd*	slope analysis	
Cortina (1995a)	XAD 2	*MDPOHEPA*[3]	*Zn, Cu, Cd*	slope and FTIR analysis	
Cortina (1995b)	XAD 2	*TOPO*[4] *and D2EHPA*	*Zn, Cu, Cd*	slope analysis	
Cortina (1994a)	Levextrel	*D2EHPA*	*Zn, Cu, Cd*	FTIR analysis	
Cortina (1994b)	XAD 2	*D2EHPA*	*Zn, Cu, Cd*	slope analysis	

[2] Di (2,4,4-trimethylpentyl)phosphoric acid (*DTMPPA*)

[3] O-methyl-dihexyl-phosphor-oxide-o-hexyl-2-ethyl phosphoric acid (*MDPOHEPA*)

[4] Tri-n-octylphosphinoxide (*TOPO*)

Literature	Resin	Ion exchanger	Ion	Equilibrium	Kinetics
Cortina (1993a,b)	XAD 2	*TMPPA*	*Zn, Cu, Cd*	slope analysis	
Cortina (1992)	XAD 2	*TMPPA*	*Zn, Cu, Cd*	selectivity	
Juang (1997)	XAD 2[5]	*D2EHPA*	*Fe, Cu, Co, Ni*		Elovich equ.
Juang (1996a,c)	XAD 4[6] XAD 16[7]	*TOA*[8]	citric acid		Elovich equ.
Juang (1996b)	XAD 2, XAD 4	*D2EHPA*	*Zn, Cu*		detailed
Juang (1995a)	XAD 2, XAD 4	*D2EHPA*	*Cu*		shrinking core model
Juang (1995b)	XAD 2, XAD 4	*D2EHPA*	*Zn*		shrinking core model
Juang (1995c)	XAD 2, XAD 4	*TOA*	citirc acid	model	
Juang (1992a)	XAD 2	*D2EHPA*	*Zn, Cu*		qualitative
Juang (1992b)	XAD 2	*D2EHPA*	*Zn*	slope analysis	
Kitazaki (1996)	XAD 7	*Cyanex 923*	Baicalein	model	qualitative
Villaescusa (1996)	XAD 2	*TOMACl*[9]	*Au*	slope analysis	
Tavlarides (1997)	ISPE[10]	*D2EHPA*	*Zn, Cd, Pb*	qualitative	qualitative
Schöneberger (2000)	DELOXAN AP II	*D2EHPA*	*Zn, HCl*	slope analysis	qualitative
Traving (2000)	XAD 4 XAD16	*TOA*	citric acid and tataric acid	slope analysis	Weisz–Prater criterion

The reactive equilibria depend on the reaction mechanism of the ion exchanger as discussed in Chapter 1. The reactive extraction of a solute, *A*, with the anion exchanger, *B*, according to the stoichiometry, *p*, is as follows:

$$p\,A + \overline{B} \leftrightarrow \overline{A_p B} \tag{4.18}$$

$$K = \frac{[\overline{A_p B}]}{[A]^p\,[\overline{B}]} \tag{4.19}$$

with $[\overline{B}] = [\overline{B_0}] - [\overline{A_p B}]$

[5] Styrol-divinylbenzol copolymer (XAD 2)
[6] Styrol-divinylbenzol copolymer (XAD 4)
[7] Divinylbenzolpolymer (XAD 16)
[8] Trioctylamine (*TOA*)
[9] Trioctylmethylammoniumchloride (*TOMACl*)
[10] Zeolite (*ISPE*)

$$K = \frac{[\overline{A_p B}]}{[A]^p \,([\overline{B_0}] - [\overline{A_p B}])} \tag{4.20}$$

$$[\overline{A_p B}] = \frac{K\;[\overline{B_0}][A]^p}{1 + K\,[A]^p} \tag{4.21}$$

The concentration of the ion exchanger can be related to any kind of adsorption isotherm. Fig. 4.12 gives the equilibrium at three different temperatures when extracting citric acid with *TOA* adsorbed on XAD 16.

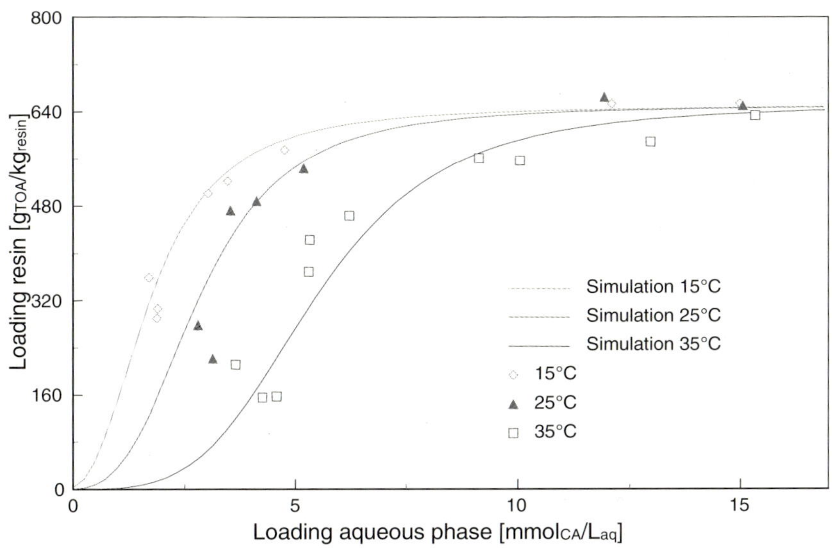

Fig. 4.12: Equilibria of the citric acid/XAD 16 system at different temperatures (initial *TOA* loading of 520 g/kg resin)

The mass transfer in the grain can be easily estimated in a stirred vessel experiment, where only resistances inside the resin prevail at high stirring rates. The use of the Weisz–Prater criterion, Ψ, helps to decide whether reaction or diffusional resistances dominate the mass transfer [Fogler (1992)]:

$$\Psi = \frac{r_{eff} \cdot L_c^2}{D_{eff} \cdot x_{aq}} \tag{4.22}$$

where r_{eff} is the initial reaction rate and D_{eff}, the effective diffusivity from experimental data. The characteristic length L_c for a spherical particle is:

$$L_c = \frac{V_P}{A_P} = \frac{R_P}{3} \tag{4.23}$$

The Weisz–Prater number is shown in Fig. 4.13 is at different temperatures

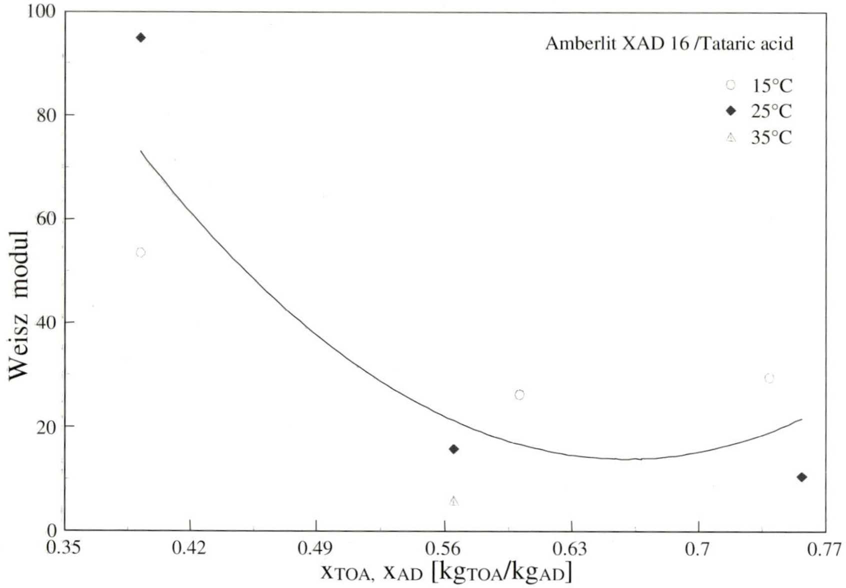

Fig. 4.13: Weisz–Prater criterion for the reactive sorption of tataric acid on XAD16

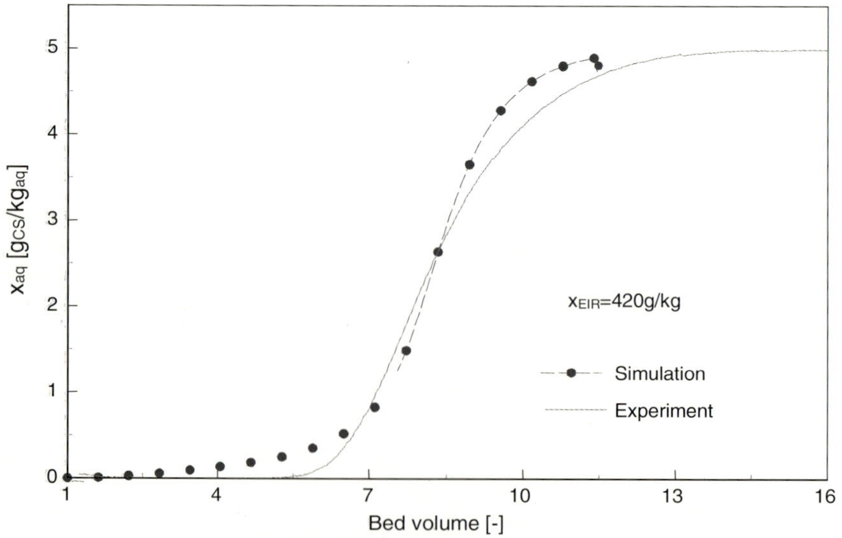

Fig. 4.14: Comparison of experimental and simulation break-through curves for citric acid/XAD 16/*TOA* ($T = 298$ K, $k_f = 3.85$ cm^{-2}/s, $D_p = 4.8\times10^{-6}$ cm^2/s, $u_D = 0.27$ mm/s, $D_{ax} = 1$ mm^2/s)

for the extraction of tataric acid with XAD 16. As can be seen, the relevant transfer mechanism is controlled by diffusion, since the critical value of $\Psi > 7$ is always fulfilled. For the even smaller pore diameter in the resin XAD 4 and for the alternatively more bulky solute, citric acid, this criterion is the more valid (not depicted).

The appropriate mass transfer information has to be implemented in a model describing a fixed bed [Traving (2000)]:

$$\frac{\partial x(z,t)_{aq}}{\partial t} = D_{ax} \frac{\partial^2 x(z,t)_{aq}}{\partial z^2} - \frac{u_0}{\varepsilon} \frac{\partial x(z,t)_{aq}}{\partial z} - \frac{3}{R_p} \varepsilon_p D_{eff} \left(\frac{\partial x(r,t)_{aq}}{\partial r} \right) \tag{4.24}$$

The first term on the right side accounts for backmixing with the axial dispersion coefficient, D_{ax}, the second is the convective term with the velocity of the continuous phase, u_0, in the bed with porosity, ε, and the last is the mass transfer into the pellet of radius, R_p, and the pellet porosity, ε_p. The simulations according to this model are given in Fig. 4.14.

For a summary to this technique see the publications listed in Table 4.1. As has been shown here, the use of dimensional numbers helps to evaluate the mass transfer regime and to define the mass balance in a fixed bed adsorber. The technique is easy to apply to different problems. However, despite the wide potential in bio- and environmental applications the use of this technique on an industrial scale has not been previously reported, providing an encouraging prospect for its future use.

Appendix A – Conversion from Molar to Molal

In electrochemistry the molality concentration scale is usually used which gives the concentration of moles solute per kg solvent. The concentration from a molar scale is as follows:

$$m_i = \frac{c_i}{\rho_L - 0.001 \sum_{i=1}^{n} M_i \, c_i} \tag{A1}$$

M_i [kg/kmol] denotes the molecular weight, ρ_L [kg/L] is the density of the solution and is usually unknown. An approximation for it is the apparent molar volume Φ_i^V, where the molar volume V_L of a solution is the result.

$$\Phi_i^V = \frac{V_L - n_w \, V_w}{n_i} \tag{A2}$$

V_w is the molar volume of water at the appropriate temperature, n_w denotes the number of water molecules and n_i that of ions. Φ_i^V is thus the volumetric difference of a solution and pure water related to one mol of an electrolyte. Masson (1929) showed Φ_i^V to be proportional to the square root of the electrolyte concentration:

$$\Phi_{M_pX_q}^V = \Phi_{M_pX_q}^0 + S_{M_pX_q} \sqrt{c_{M_pX_q}} \tag{A3}$$

Φ^0 is the apparent molar volume of an electrolyte M_pX_q at infinite dilution. For a single electrolyte equ. (A3) is a good approximation. For mixtures the following is valid:

$$\Phi_{M_pX_q}^0 = p\,\Phi_M^0 + q\,\Phi_X^0 \tag{A4}$$

$$S_{M_pX_q} = p\,S_M + q\,S_X \tag{A5}$$

$$\Phi_{M_pX_q}^V = p\,\Phi_M^V + q\,\Phi_X^V \tag{A6}$$

and thus for a single ion:

$$\Phi_i^V = \Phi_i^0 + S_i \sqrt{c_{M_pX_q}} \tag{A7}$$

However, Φ_i^V can therefore be calculated for one salt at its concentration c. Equ. (A7) can be rewritten:

$$\Phi_i^V = \Phi_i^0 + \frac{S_i}{\sqrt{\left(z_M^2\ p + z_X^2\ q\right)/2}} \sqrt{I^c} \qquad \text{(A8)}$$

where the ionic strength I^c for a salt M_pX_q on a molar basis is:

$$I^c = \tfrac{1}{2}\left(z_M^2\ c_M + z_X^2\ c_x\right) = \tfrac{1}{2}\left(z_M^2\ p\ c_{M_pX_q} + z_X^2\ q\ c_{M_pX_q}\right) \qquad \text{(A9)}$$

In order to calculate the apparent molar volume of a multicomponent solution, it is assumed to be only a function of I^c at a given temperature. This relies on experimental results that negligible volume change will occur when mixing solutions of equal ionic strength [Young & Smith (1954)]. Equ. (A9) is thus valid in a salt mixture, where

$$I^c = \tfrac{1}{2}\sum_{i=1}^{n} z_i^2\ c_i \qquad \text{(A10)}$$

In order to calibrate the proportionality factor S_i the following conversion was made:

$$\Phi_{H^+}^0 = 0 \text{ and } S_{H^+} = 0 \qquad \text{(A11)}$$

HCl in water serves as an example for this where the concentration dependency of the density must be known. According to Harned and Owen (1958):

$$\Phi_i^V = \frac{1000}{c_i\ \rho_w}\left(\rho_w - \rho_L\right) + \frac{M_i}{\rho_w} \quad [\text{cm}^3/\text{mol}] \qquad \text{(A12)}$$

Plotting Φ_i^V versus the square root of the concentration yields Φ^0 and S. With the zero contribution according equ. (A12) the following is valid:

$$\Phi_{HCl}^0 = \Phi_{Cl^-}^0 \quad \text{and} \quad S_{HCl} = S_{Cl^-} \qquad \text{(A13)}$$

With the mixing rules (equs. (A4) and (A5)) Φ and S of any other cation can be estimated if density data for the chloride salt are available. Equ. (A12) can be rewritten and for a single ion is thus:

$$\left(\rho_w - \rho_L\right) = \frac{c_i\left(\rho_w\ \Phi_i^V - M_i\right)}{1000} \qquad \text{(A14)}$$

and accordingly for n ions:

$$\left(\rho_w - \rho_L\right) = \frac{1}{1000}\sum_{i=1}^{n}\left(\rho_w\ \Phi_i^V - M_i\right)c_i \qquad \text{(A15)}$$

and finally:

$$\rho_L = \rho_w - \frac{1}{1000}\sum_{i=1}^{n}\left(\rho_w \Phi_i^V - M_i\right)c_i \tag{A16}$$

A compendium of Masson parameters can be found from Milero (1972) and Harned and Owen (1958).

In order to convert the equilibrium constant of the law of mass action we rely on the following limits, where v denotes the stoichiometric factor:

$$\lim_{\substack{c_i \to 0 \\ i=1\ldots n}} K^c = \prod_{i=1}^{n}(c_i)^{v_i} \tag{A17}$$

$$\lim_{\substack{m_i \to 0 \\ i=1\ldots n}} K^m = \prod_{i=1}^{n}(m_i)^{v_i} \tag{A18}$$

and

$$\lim_{c_i \to 0}\left(\frac{c_i}{m_i}\right) = \rho \tag{A19}$$

where ρ [kg/L] is the density of the solvent. For a homogeneous reaction there is:

$$K^c = K^m\,(\rho)^{\sum v_i} \tag{A20}$$

and for a heterogeneous reaction we have to consider the density of the diluent (ρ_{DIL}) and of water (ρ_w):

$$K^c = K^m\,(\rho_{DIL})^{\left(\sum v_i\right)_{DIL}}\,(\rho_w)^{\left(\sum v_i\right)_w} \tag{A21}$$

Appendix B – Activity Coefficient Conversion

The chemical potential of a binary electrolyte M_pX_q reads as:

$$\mu_{MX} = p\left[\mu_M^{\oplus} + RT\ln\left(\gamma_M^m\, m_M\right)\right] + q\left[\mu_X^{\oplus} + RT\ln\left(\gamma_x^m\, m_X\right)\right] \tag{B1}$$

with $m_M = p\,m$ and $m_x = q\,m$ is:

$$\mu_{MX} = m_{MX}^{\oplus} + RT\ln\left(m^{p+q}\, p^p\, q^q\right) + RT\ln\left(\gamma_M{}^p\, \gamma_X{}^q\right) \tag{B2}$$

with

$$m_{MX}^{\oplus} = p\,m_M^{\oplus} + q\,m_X^{\oplus} \tag{B3}$$

Since there exists no solution of single cations or anions what can be measured is a mean or average activity coefficient:

$$\gamma^{\pm} = \left(\gamma_M{}^p\, \gamma_X{}^q\right)^{\frac{1}{p+q}} = \left(\gamma_M{}^p\, \gamma_X{}^q\right)^{\frac{1}{s}} \tag{B4}$$

and

$$m^{\pm} = (m_M^p\, m_X^q)^{\frac{1}{s}} \tag{B5}$$

with $s=p+q$. The expression for the chemical potential of a dissolved salt is then:

$$\mu_{MX} = \mu_{MX}^{\oplus} + RT\ln(m^{\pm}\,\gamma^{\pm})^s \tag{B6}$$

In real solutions, the activity of water is quite close to one as the dilution ratio of the salts increases. In order to represent accurately the activity of water, several significant digits would be required. To avoid this problem, in many compilations of data it is common to tabulate data in terms of the osmotic coefficient ϕ_w. It is derived considering non-idealities in the van't Hoff equation with membrane processes and is defined as [Horvarth (1985)]:

$$\phi_w = -\frac{1000}{M_w \sum_{i=1}^{n} m_i}\ln a_w \tag{B7}$$

It can be derived from the Gibbs excess enthalpy as to

$$\phi_w - 1 = -\frac{\left(\partial G^{EX} \big/ \partial n_w\right)_{T,P,n_j \neq n_w}}{RT \sum_{i=1}^{n} m_i} \tag{B8}$$

The conversion of activity coefficients of electrolytes from one scale to another is presented by Robinson and Stokes (1965) as to:

- mole fraction to molal:

$$\gamma_{\pm}^{(x)} = \left(1 + 0.001\, M_W\ s\ m\right)\gamma_{\pm} \tag{B9}$$

- mole fraction to molar:

$$\gamma_{\pm}^{(x)} = \left(\frac{\rho_L + 0.001\, c\left(s\, M_W - M_i\right)}{\rho_w}\right)\gamma_{\pm}^{(c)} \tag{B10}$$

- molal to molar:

$$\gamma_{\pm} = \left(\frac{\rho_L - 0.001\, c\, M_i}{\rho_w}\right)\gamma_{\pm}^{(c)} = \left(\frac{c}{m\,\rho_w}\right)\gamma_{\pm}^{(c)} \tag{B11}$$

- molar to molal:

$$\gamma_{\pm}^{(c)} = \left(1 + 0.001\, m\, M_i\right)\frac{\rho_w}{\rho_L}\gamma_{\pm} = \left(\frac{m\,\rho_w}{c}\right)\gamma_{\pm} \tag{B12}$$

For non-electrolytes a similar procedure can be derived on the basis of the chemical potential on molar scale:

$$\mu_i = \mu_i^0(c) + RT\ \ln\left(\frac{c_i\,\gamma_i^{(c)}}{c^0}\right) \quad \text{with } c^0 = 1\ \text{mol/L} \tag{B13}$$

and on mole fraction scale:

$$\mu_i = \mu_i^0 + RT\ \ln\left(x_i\ \gamma_i^{(x)}\right) = \mu_i^0 + RT\ \ln\left(a_i\right) \tag{B14}$$

thus

$$\ln\frac{\gamma_i^{(c)}}{\gamma_i^{(x)}} = \ln\left(\frac{x_i}{c_i}\right) + \frac{\mu_i^0 - \mu_i^0(c)}{RT} \tag{B15}$$

The last term in this equation is independent of concentrations and is thus defined considering the following limit:

$$\lim_{c_i \to 0} \gamma_i^{(c)} = 1 \tag{B16}$$

with $c \to 0$:

$$\lim_{c_i \to 0}\left(\ln\frac{\gamma_i^{(c)}}{\gamma_i^{(x)}} + \ln\left(\frac{c_i}{x_i}\right)\right) = \ln\left(\frac{1}{\gamma_i^{\infty}}\right) + \ln(c_s) = \frac{\mu_i^0 - \mu_i^0(c)}{RT} \qquad \text{(B17)}$$

The ratio c_i/x_i at $c \to 0$ approaches the molar concentration of the solvent, c_s. It follows:

$$\gamma_i^{(c)} = \gamma_i^{(x)}\left(\frac{x_i}{c_i}\right)\left(\frac{c_s}{\gamma_i^{\infty}}\right) \qquad \text{(B18)}$$

The same procedure holds for the molality scale conversion with reference (see equ. (B13)) $c^0 = 1$ kg/L, where m_s is the molal solvent concentration:

$$\gamma_i^{(m)} = \gamma_i^{(x)}\left(\frac{x_i}{c_i}\right)\left(\frac{m_s}{\gamma_i^{\infty}}\right) \qquad \text{(B19)}$$

Appendix C – Operation and Design of a Sieve Tray

A large part of industrial extraction columns are made up of sieve trays. They are either used in a non-agitated mode or more often in a reciprocating mode. A rapid reciprocating motion imparted to the liquid in a tower results in an improved mass transfer. This action can be accomplished without moving parts or bearings coming into contact with the liquid phase and thus has found an application for handling hazardous and corrosive liquids in the chemical and nuclear industry. However, an alternative to the pulsation of the liquid is by a reciprocating motion of the plates [Lo et al. (1983)].

The sieve tray column (see Fig. C-1) was first patented by Laird (1919) and can be designed either to disperse the heavy or the light phase. The continuous phase passes across each tray and proceeds to the next one through a downcomer or a riser. The dispersed phase is trapped and coalesced at each tray in a layer and redispersed. The axial dispersion is thus limited between two trays within these coalesced layers. The repeated stagewise dispersion and thus surface renewal have generally improved the efficiency in comparison to other types of non-agitated towers.

Due to the complex hydrodynamics multi-pass trays are not usual and the column diameter is usually smaller than 3.66 m. The tray deviations from horizontal should be less than ±1mm, which also limits column diameters. In large trays the two halves of the tray are separately removable and the tray spacing should be at least 0.4 m in order to provide entry ports for cleaning and installation. The tray is sealed with a valance or a spiral spring in a cage, since otherwise it gives rise to emulsion formation, and in contrast to absorption/distillation there is no weir. The clearance under the downcomer is usually a quarter of the tray spacing. The hole diameters are much smaller than in gas–liquid systems with 2–8 mm (2 mm is a fabrication limit) and are usually set in triangular (square) arrays on about 16mm centres. There appears to be little effect of the hole size on the extraction rate, but with systems of higher interfacial tension, smaller holes should be favoured. They occupy from 5 to 63% of the available tray area (default 15%). The velocities through the holes should be kept between 0.15 and 0.3 m/s. If the downcomer is equipped with coalescence aids, the downcomer section must be made correspondingly larger. Further details can also be found in Treybal (1963), Skelland and Conger (1973), Pilhofer and Mewes (1979), Cavers (1983), Humphrey et al. (1994), Robbins and Cusack (1997) and Mewes and Pilhofer (1979). In the following the design and the operation of a sieve tray column is discussed (http://www.uni-kl.de./LS-Bart/DAE).

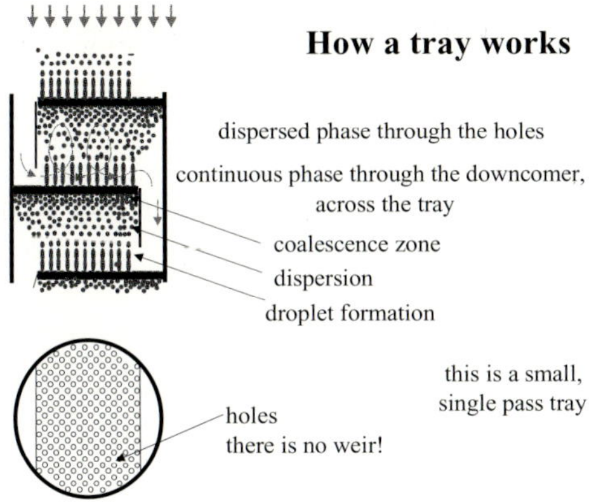

Fig. C-1: The sieve tray

C.1 Operating and Design Variables

Once a sieve tray has been constructed, we can still change conditions on the tray via operation variables. Here, we only consider the internal flow rates of the heavy and light liquid as operating variables. We do not consider changes in temperature but have to balance the pressure drop at each tray for functionality. Also in contrast to gas–liquid systems, the hold-up of the dispersed phase ϕ_d plays a decisive role for flooding.

The design variables determine the shape and geometry of the sieve tray. Important horizontal variables are:

- the diameter d_h of the holes,
- the area A_h of the holes,
- the area A_d of the downcomers or risers,
- the total area A_T of the column.

Important vertical dimensions are

- the tray spacing or height H_T,
- the height of the coalescence layer h_c,
  Summarising:
- the operating variables are the continuous, w_c, and dispersed phase velocity, w_d and the hold-up ϕ_d, and
- the design variables are A_T, A_d, A_h, H_T, h_c and d_h.

C.2
Operating Limits

The total volumetric flow rates of continuous and dispersed liquid though the column are equal to Q_c and Q_d. Together with the tray area A_T they define the superficial velocities [m/s]:

$$w_d = \frac{Q_d}{A_T} \tag{C1}$$

$$w_c = \frac{Q_c}{A_T} \tag{C2}$$

The operating range of a tray is a limited area in a diagram of the superficial velocities (see Fig. C-2). Outside this range, the performance of a tray falls to unacceptable values. Until flooding occurs there is only one flow regime inside these boundaries, where phase inversion will occur. The limits are as such:

- Entrainment:
  Liquid in the downcomer entrains small dispersed phase droplets ($w_{entrain}$), usually at high continuous phase flows.
- Flooding:
  With small droplets and high hold-up the droplet rise approaches zero or at extremely high hold-up phase inversion will occur (w_{flood}).
- Inactive holes:
  The Weber number in the holes should exceed two, to ensure all holes produce drops (w_{min}).
- Min. coalesced height:
  A certain layer of dispersed phase is necessary to have all perforations working unless the tray is exactly level (w_{hcmin}).
- Max. coalesced height:
  It equals the downcomer height (z_d), which is usually three quarters of the tray spacing (H_T).

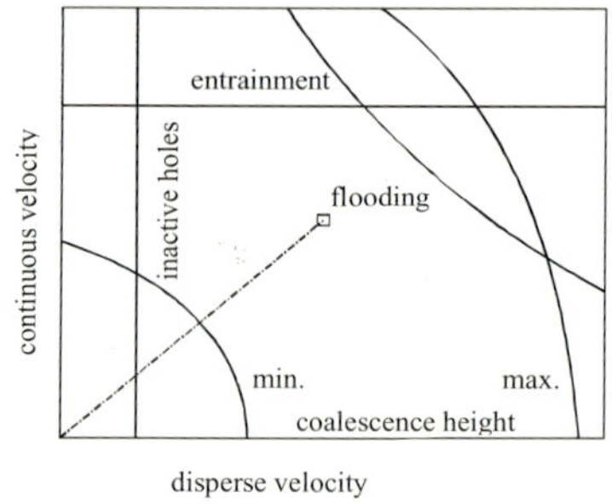

Fig. C-2: Operating limits

C.3
The Many Variables

Many geometrical and physical factors influence the limits of the operating range of the tray. Experience shows that the most important factors are:

- geometric ones like d_h, A_h, A_T, H_T,
- the superficial velocities w_c and w_d,
- the height of the coalesced dispersed phase h_c,
- the densities, ρ_c, ρ_d, and viscosities, η_c, η_d and the interficial tension σ,
- the gravitational acceleration g.

C.4
Boundaries of the Operating Range and Design Variables

In this section we derive relations for the different limits of the operating range of a sieve tray. This will be plotted in an operating diagram (see Fig. C-3). It is not possible to avoid empirical relations and rules. However, we shall try to explain their behaviour when they are needed. Otherwise, we will use simple physical models of flow phenomena. These are not always accurate, but they give the designer a good feel for the relation between the many different variables involved in a tray design.

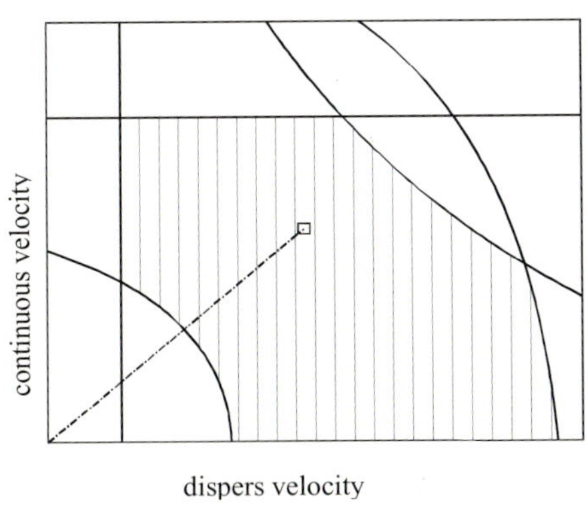

Fig. C-3: Operating diagram

C.4.1
Inactive Holes

The most decisive factor in a sieve tray design is in a proper choice of the hole diameter, d_h. The hole diameter d_h is set by default to:

$$\chi = \sqrt{\frac{\sigma}{\Delta\rho g}} \tag{C3}$$

$$d_h = 1.8\chi \tag{C4}$$

but d_h is limited (if supplied) by:

$$0.5\chi < d_h < \pi\chi \tag{C5}$$

and the practical limits (over-riding):

$$3\text{mm} < d_h < 8\text{mm} \tag{C6}$$

The hole velocity is computed with the Eötvös and the Weber number:

$$E\ddot{o} = \frac{\Delta\rho\ g\ d_h^2}{\sigma} \tag{C7}$$

$$We = 4.33 E\ddot{o}^{-0.26} \tag{C8}$$

$$U_h = \sqrt{\frac{We\ \sigma}{\rho_d\ d_h}} \tag{C9}$$

All perforations in a sieve tray will operate [Ruff (1974)] if the Weber number in the hole exceeds two (which is about 0.15m/s hole velocity). The default value of We=2 (w_{min}) appears as a vertical line in the operating diagram,Fig. C-4.

The hole diameter also determines the resulting droplet diameter. For $E\ddot{o}$ is less than 0.4, the Sauter mean droplet diameter is computed by:

$$d_p = E\ddot{o}^{-0.4}\left(2.13\left(\frac{\Delta\rho}{\rho_d}\right)^{0.67} + \exp(-0.13Fr)\right)d_h \tag{C10}$$

otherwise

$$d_p = E\ddot{o}^{-0.42}\left(1.24 + \exp\left(-Fr^{0.42}\right)\right)d_h \tag{C11}$$

An alternative approach is with:

$$d_p = d_h\ 2.1\times 10^{-(0.094\ E\ddot{o})} \tag{C12}$$

The Froude number is computed from

$$Fr = \frac{U_h^2}{g\cdot d_h} \tag{C13}$$

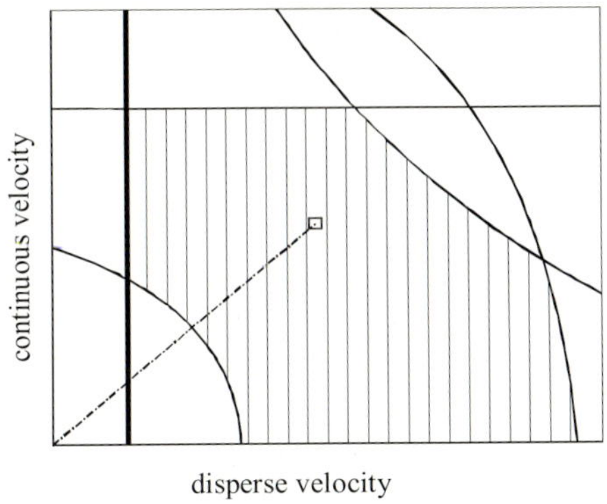

Fig. C-4: Inactive holes

C.4.2 Entrainment

The downcomer velocity can be computed if a minimum droplet diameter d_{min} is assumed which will not be entrained. The downcomer velocity of the continuous phase U_c is:

$$U_c = 0.249 d_{min} \left(\frac{(g\Delta\rho)^2}{\rho_c \eta_c} \right)^{0.33} \tag{C14}$$

This droplet diameter is taken to be 0.7 mm. This is depicted as the only horizontal line ($w_{entrain}$) in the operating diagram (see Fig. C-5).

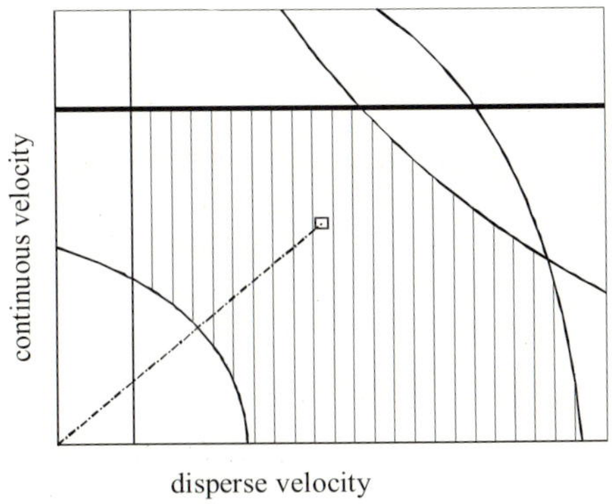

Fig. C-5: Downcomer velocity limit

C.4.3 Column Diameter

The continuous phase throughput Q_c and the dispersed phase throughput Q_d determine the total tray area and thus the column diameter. This geometrical design variables can then be derived as follows.

The hole area depends on the dispersed phase throughput Q_d as to:

$$A_h = \frac{Q_d}{U_h} \tag{C15}$$

The ratio of the hole area over the active area (free area ratio, *f*) is limited between 1 and 63% (default is 15%).

$$A_b = A_h / f \tag{C16}$$

The hole pitch can be computed if the hole diameter and free area ratio are known. With U_c known we can compute the downcomer area:

$$A_d = Q_c / U_c \tag{C17}$$

The total area is equal to two downcomer areas plus the active area and 0.5% area for support, etc.:

$$A_T = (A_b + 2A_d)/0.995 \tag{C18}$$

With the total tray area known the column diameter can be computed.

The downcomer may either be fabricated likewise in distillation (but without weir) or simply as the downcomer or riser tube. The plate is not perforated beneath a downcomer at the downspot, which introduces the factor two in the above equation.

C.4.4 Height of the Coalescence Layer

The minimum height (w_{hcmin}) is 2 cm for small columns and 5 cm for big ones (diameter >1 m) to have all perforations working unless the tray is not exactly level. The maximum height (w_{hcmax}) is as long as the downcomer length, which is usually three-quarters of the tray spacing.

The pressure drop is caused by two phenomena:

- the loss of kinetic energy of the dispersed phase when it leaves the holes and
- the head of the coalesced liquid dispersion on the tray

The downcomer delivers a counter pressure consisting of two contributions:

- the static pressure due to the liquid in the downcomer and
- a negative contribution due to kinetic energy losses of the continuous phase leaving the downcomer

At the effective coalescence height, which equals the minimum downcomer height, the inside of the downcomer is full of continuous phase and the outside is surrounded by the coalesced dispersed phase. The pressure balance then reads as [Mewes & Pilhofer (1978), Mewes & Kunkel (1977)] (see Fig. C-6):

$$\Delta p(\text{Downcomer}) = \Delta p(\text{Tray})$$

$$\Delta p_D = \Delta p_T + \rho_d\, g\, h_d + \left[\rho_d\, \phi_d + \rho_c\,(1-\phi_d)\right] g\,(z_d - h_c) \tag{C19}$$

where [Pilhofer & Goedl (1977)]:

$$\Delta p_T = 0.5 \frac{\rho_d\, U_h^2}{\sqrt{1-\dfrac{0.71}{lg(Re_h)}}} + 3.2\ \text{Eö}^{0.2}\ \frac{\sigma}{d_h} \tag{C20}$$

$$\Delta p_D = 2.47\ \rho_c \left(\frac{Q_c}{A_d}\right)^2 \tag{C21}$$

This effective coalescence height must be within the limits of the minimum and maximum height as given in Fig. C-7.

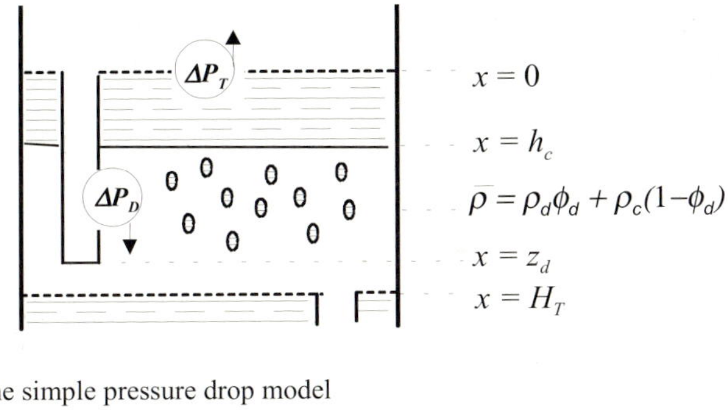

Fig. C-6: Pressure terms in the simple pressure drop model

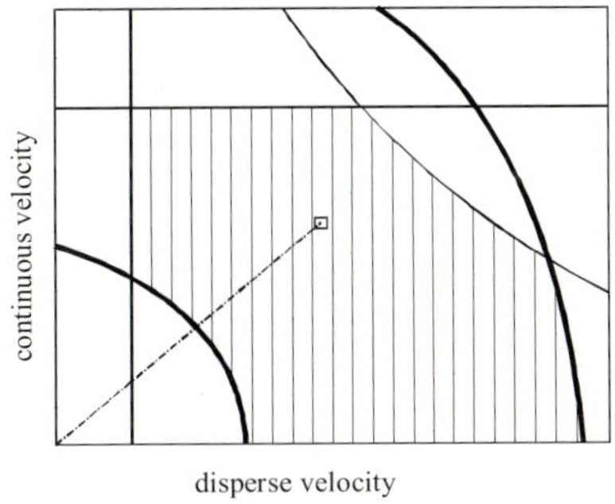

Fig. C-7: Minimum and maximum and effective coalescence height

C.4.5
Slip Velocity and Flooding

According to the two-layer model of Gayler et al.(1953) in a countercurrent system the relative or slip velocity between the phases is:

$$V_s = \frac{w_d}{\phi_d} + \frac{w_c}{(1-\phi_d)} \tag{C22}$$

When the hold-up, ϕ_d, approaches zero, this is a single droplet rising velocity or the characteristic velocity. However, the slip velocity of a dispersed droplet assembly is modified by physico-chemical data, by the drop size and the dimensionless Archimedes number (Ar), Hadamard–Rybcynski number (K_{HR}), the fluid (K_F) and the Reynolds number (Re):

$$V_s = f(\mathrm{Ar}, \mathrm{K_F}, \mathrm{K_{HR}}, \mathrm{Re}, \phi_d) \tag{C23}$$

Equs. (C22) and (C23) are depicted in Fig. C-8. Point A can be determined, when starting the iteration procedure at low hold-up (e.g. 0.1%).

Flooding occurs, when both curves meet at the maximum hold-up value. This is after Thornton (1957), where the flooding velocities are derived as follows:

$$\left(\frac{\mathrm{d}w_c}{\mathrm{d}\phi_d}\right) = \left(\frac{\mathrm{d}w_d}{\mathrm{d}\phi_d}\right) = 0 \tag{C24}$$

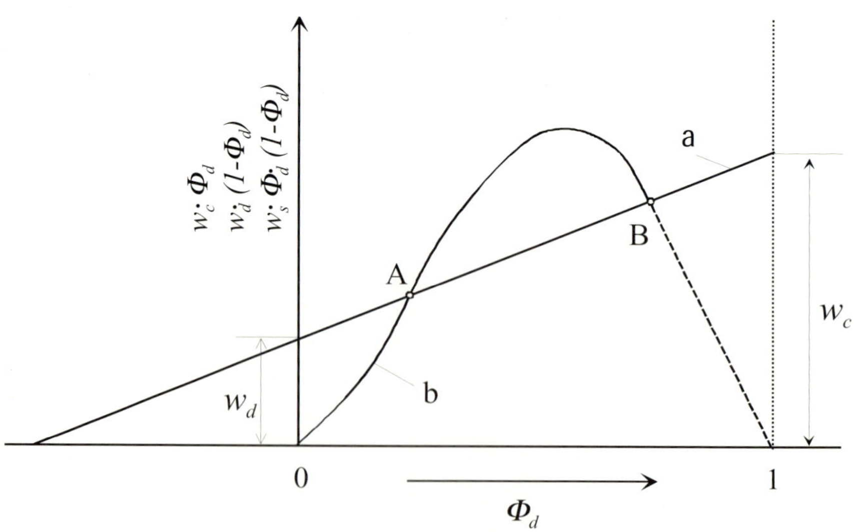

Curve a: $V_s \cdot \phi_d (1-\phi_d) = w_d + \phi_d (w_c - w_d)$

Curve b: $V_s \cdot \phi_d (1-\phi_d) = f(Ar, K_F, K_{HR}, Re, \phi_d)$

Fig. C-8: Slip velocity

According to the velocity correlations of droplet swarms the following is valid [Pilhofer & Mewes (1979)]:

$$Re_s = \frac{3\, zq^2\, \phi_d}{\xi\, q^3 (1-\phi_d)} \left[\left(1 + \frac{\xi}{54} \frac{q^3}{(zq^2)^2} \frac{(1-\phi_d)^3}{\phi_d^2} Ar \right)^{\frac{1}{2}} - 1 \right] \tag{C25}$$

where

$$\begin{aligned} q^3 &= \left(\frac{\phi_d}{1-\phi_d} \right)^{0.45} \frac{5}{K_{HR}^{3/2}} \\ zq^2 &= \frac{1}{K_{HR}} \left(\frac{1-\phi_d}{A\phi_d} \right) \exp\left(\frac{B\,\phi_d}{1-0.61\phi_d} \right) \\ \xi &= \left(\frac{1}{6} \frac{Ar}{Re_\infty} - \frac{3}{K_{HR}} \right) \frac{1}{Re_\infty} \\ Re_\infty &= K_F^{\;0.15} \left(Ar^{0.523}\, K_F^{\;-0.114} - 0.75 \right) \end{aligned} \tag{C26}$$

Equ. (C25) is valid in technically relevant regions ($0.06 < \phi_d < 0.55$) and at dimensionless drop diameter $Ar > 1$ for circulating drops ($A = 2$; $B = 2.5$). For a higher hold-up ($0.55 < \phi_d < 0.74$) there is a change in zq^2 as $A = 0.45$ and $B = 0.44$. For oscillating drops ζ in equ. (C25) changes to:

$$\begin{aligned} Ar \le Ar_G &= 394 \cdot K_F^{\;0.275} \\ \xi &= \left(\frac{1}{6} \frac{Ar}{Re_\infty} - \frac{3}{K_{HR}} \right) \frac{1}{Re_\infty} C \end{aligned} \tag{C27}$$

$$\begin{aligned} C &= 1 \\ Re_\infty &= K_F^{\;0.15} \left(Ar^{0.523} \cdot K_F^{\;-0.114} - 0.75 \right) \end{aligned} \tag{C28}$$

and

$Ar > Ar_G$

$$Re_\infty = K_F^{\;0.15} \left(4.178 Ar^{0.281}\, K_F^{\;-0.0773} - 0.75 \right) \tag{C29}$$

and for $\eta_F > \eta_p$ here C again equals one whereas for $\eta_F < n_p$:

$$C = \left(1 - \phi_p^{\;1.29} \left(\frac{Ar - Ar_G}{Ar_G} \right)^{-1.74} \right) \tag{C30}$$

The deviation of equs. (C22) and (C25) according to equ. (C24) yields the flooding curve in the operating diagram (Fig. C-9).

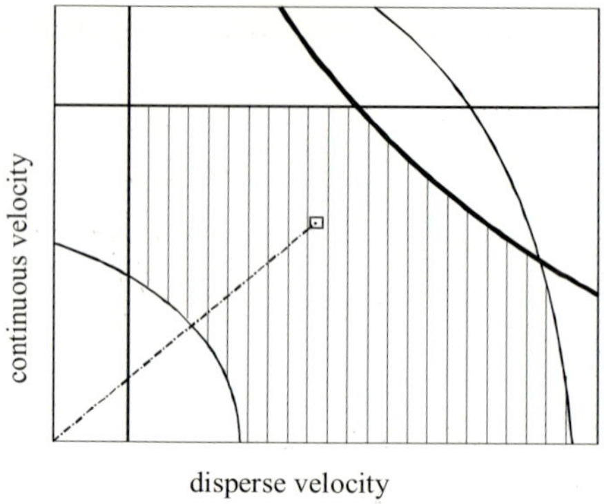

Fig. C-9: Flooding curve

C.4.6 Tray Operation

In this manual we have presented the hydrodynamic restrictions of a sieve tray. It is not difficult to check whether a tray can handle certain dispersed and continuous phase flows and we can also see the location of the operating point with respect to operating limits. We can thus check whether the tray is suitable for a certain operation.

Input

We need the following data:

- the continuous phase flow and properties: M_c, Q_c, ρ_c, η_c,
- the dispersed phase flow and properties M_d, Q_d, ρ_d, η_d,
- the interfacial tension and the gravitational acceleration, g,
- the tray spacing, H_T,
- the downcomer height, z_d.
- the column diameter, D_k,
- the downcomer diameter, d_d,
- the area fraction, f, occupied by holes (perforations).

From these data, we calculate the tray, downcomer, active and hole areas, the number of holes and the pitch:

$$A_T = \frac{\pi}{4} D_k^2 \tag{C31}$$

$$A_d = \frac{\pi}{4} d_d^2 \tag{C32}$$

$$A_b = A_T\ 0.995 - 2A_d \tag{C33}$$

$$A_h = f\ A_b \tag{C34}$$

$$N_h = A_h \Big/ \left(\frac{\pi}{4} d_h^{\,2} \right) \tag{C35}$$

$$pitch = d_h (0.907/f)^{1/2} \tag{C36}$$

The Operating Point

We need the operating point in terms of superficial velocities:

$$w_c = \frac{M_c}{A_T} \frac{Q_c}{\rho_c} \tag{C37}$$

$$w_c = \frac{M_d}{A_T} \frac{Q_d}{\rho_d} \tag{C38}$$

We plot these in a diagram of w_c versus w_d.

To calculate the operating conditions we first have to calculate the droplet diameter:

$$d_p = d_h \; 2.1 \times 10^{-(0.094\,\text{Eö})} \tag{C39}$$

The iterative calculation of hold-up and slip velocity at the operating point:

$$V_s = \frac{w_d}{\phi_d} + \frac{w_c}{(1-\phi_d)} \tag{C40}$$

$$V_s = f\{Ar, K_F, K_{HR}, Re, \phi_d\} \tag{C41}$$

and the hole velocity at the operating point:

$$U_h = \sqrt{\frac{We\,\sigma}{\rho_d\, d_h}} \tag{C42}$$

The Operating Range

The minimum hole velocity, for $\text{We} = 2$, is:

$$U_h = \sqrt{\frac{We\,\sigma}{\rho_d\, d_h}} \tag{C43}$$

The maximum allowable downcomer velocity is produced with the smallest entrained droplet (e.g. diameter=0.7mm):

$$U_c = 0.249 d_{min} \left(\frac{(g\Delta\rho)^2}{\rho_c \eta_c} \right)^{0.33} \tag{C44}$$

The flooding line is obtained when applying equ. (C41):

$$\frac{\partial w_c}{\partial \phi_d} = 0 \tag{C45}$$

with

$$V_s = f\{Ar, K_F, K_{HR}, Re, \phi_d\} \tag{C46}$$

The limiting lines due the minimal and maximal height of the coalescence layer are calculated due to the pressure balance:

$$h_c\ g\ \rho_c - \Delta p_D = \Delta p_T + h_c\ g\ \rho_d \tag{C47}$$

with a default $h_{c,\min}$=2cm (5cm for large trays) and $h_{c,\max}$ equals the downcomer length.

We plot all these lines and see whether the operating point falls within the operating range (see Fig. C-10).

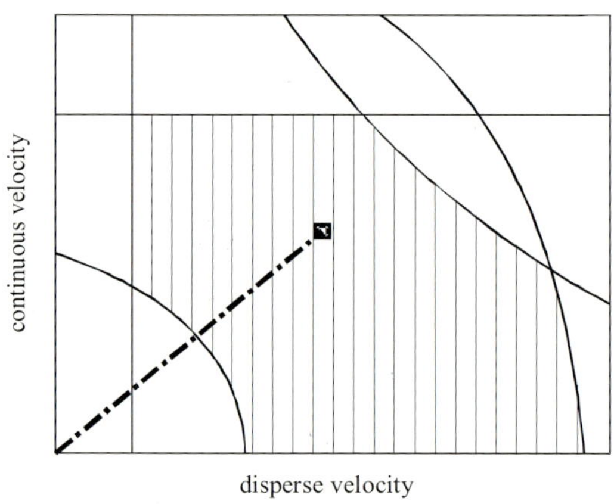

Fig. C-10: Operating point

C.4.7 Design a Sieve Tray

Making a design, is more difficult than checking whether a tray can handle a certain operation. The problem is that there is any number of possible designs. A design has to comply with several criteria, which are often conflicting:

- the tray (or the whole separation system) must be cheap,
- there should be a small "safety" margin to increase the throughput or allow for design inaccuracies,
- we should also be able to operate the tray below its design point; the tray should be flexible or have a reasonable "turn down ratio",
- the pressure drop should be low (at least in low pressure columns),

- the tray should withstand normal mechanical forces,
- it should not foul or corrode and
- it should be easy to construct and maintain.

Making a good choice in the design variables means thinking, compromising, investigating alternatives … well let us say, good engineering. The model given here can help to do this. In this final chapter, we present an algorithm, which at least gives reasonable designs.

Input

We need the following data:

- the continuous phase flow and properties: M_c, Q_c, ρ_c, η_c,
- the dispersed phase flow and properties M_d, Q_d, ρ_d, η_d,
- the interfacial tension and the gravitational acceleration, g,
- the tray spacing, H_T,
- the downcomer height, z_d.

The flows and their properties follow from design calculations. We will not discuss them here.

Tray Height and Downcomer Length

The tray height is quite an important parameter. The spacing between trays for large columns is up to 0.65 m (default is 0.4 m) providing entry ports for cleaning and installation. The tray height for small columns is between 0.05 and 0.15 m and should at least be twice as much as the height of the coalescencing zone, h_c. The default for small columns is 0.1 m, since higher clearances give rise to an increased back mixing. The downcomer length is set usually to three-quarters of the tray spacing.

$$H_T = 0.1 \text{ m}$$

$$z_d = 0.075 \text{ m}$$

Hole Diameter and Active Area

The selection of the hole diameter is between 0.002 and 0.008 m. The hole diameter d_h is set by default to:

$$\chi = \sqrt{\frac{\sigma}{\Delta\rho g}} \tag{C48}$$

$$d_h = 1.8\chi \tag{C49}$$

but d_h is limited (if supplied) by:

$$0.5\chi < d_h < \pi\chi \tag{C50}$$

and the practical limits (overriding):

$$3 \text{ mm} < d_h < 8 \text{ mm} \tag{C51}$$

The hole velocity is computed with the Eötvös and the Weber numbers:

$$Eö = \frac{\Delta\rho \; g \; d_h^2}{\sigma} \tag{C52}$$

$$We = 4.33 Eö^{-0.26} \tag{C53}$$

$$U_h = \sqrt{\frac{We \; \sigma}{\rho_d \; d_h}} \tag{C54}$$

If Eö is less than 0.4, the Sauter mean droplet diameter is computed by:

$$d_p = Eö^{-0.4} \left(2.13 \left(\frac{\Delta\rho}{\rho_d} \right)^{0.67} + \exp(-0.13 Fr) \right) d_h \tag{C55}$$

otherwise

$$d_p = Eö^{-0.42} \left(1.24 + \exp\left(-Fr^{0.42} \right) \right) d_h \tag{C56}$$

An alternative approach is with:

$$d_p = d_h \; 2.1 \times 10^{-(0.094 \; Eö)} \tag{C57}$$

The Froude number is computed from

$$Fr = \frac{U_h^2}{g \; d_h} \tag{C58}$$

The perforation area is:

$$A_h = \frac{Q_d}{U_h} \tag{C59}$$

and the number of holes:

$$N_h = A_h / (\frac{\pi}{4} d_h^2) \tag{C60}$$

The active plate area required to provide N_h perforations, for holes located at the vertices of equilateral triangles is:

$$A_b = \frac{N_h \; \pi \; (PITCH)^2}{3.62} \tag{C61}$$

and for the holes on a square pitch:

$$A_b = \frac{N_h \; \pi \; (PITCH)^2}{3.14} \tag{C62}$$

Typically a pitch of about 12–20 mm is used [Cavers (1983)].

Downcomer Area

The downcomer velocity can be computed if a minimum droplet diameter d_{min} is assumed which will not be entrained. The downcomer velocity equals the velocity of the continuous phase U_c:

$$U_c = 0.249 d_{min} \left(\frac{(g\Delta\rho)^2}{\rho_c \eta_c} \right)^{0.33} \quad \text{(C63)}$$

This droplet diameter is taken to be e.g. 0.7 mm. With U_c known we can compute the downcomer area:

$$A_d = Q_c / U_c \quad \text{(C64)}$$

The downcomer diameter is then:

$$d_d = \sqrt{A_d \, 4/\pi} \quad \text{(C65)}$$

Tray Area

The total tray area is equal to two downcomer areas plus the active area and 0.5% area for support:

$$A_T = (A_b + 2A_d)/0.995 \quad \text{(C66)}$$

The diameter of the column is then:

$$D = \sqrt{\frac{\pi}{4} A_T} \quad \text{(C67)}$$

To control the design, some empirical rules should be kept in mind. The ratio of the hole area over the active area (free area ratio, f) is limited to between 1 and 63% (and is usually smaller than 20%):

$$f = A_h / A_b \quad \text{(C68)}$$

The height of the tray spacing should be at a minimum twice the effective coalescing height, h_c, which is derived from the tray's pressure balance.

$$h_c \, g \, \rho_c - \Delta p_D = \Delta p_T + h_c \, g \, \rho_d \quad \text{(C69)}$$

Finally, the operating point should be lower than the flooding point. At the flooding points curve a and curve b in Fig. C-8 should meet at the maximum:

$$\frac{\partial}{\partial \phi_d}\left(w_d + \phi_d (w_c - w_d)\right) = \frac{\partial}{\partial \phi_d}\left(f\{Ar, K_F, K_{HR}, Re, \phi_d\}\right) = 0 \quad \text{(C70)}$$

Appendix D – The LAP Model for Multicomponent Mixtures

The extraction of HCl with TOA according to Danesi et al. (1987) is:

$$TOA_{ad} + H^{+} \leftrightarrow TOAH^{+}_{ad} \tag{D1}$$

$$TOAH^{+}_{ad} + Cl^{-} \leftrightarrow TOAHCl_{ad} \tag{D2}$$

$$TOAHCl_{ad} + TOA \leftrightarrow TOAHCl + TOA_{ad} \tag{D3}$$

The individual kinetic rate equations are then:

$$R_1 = k_1\,[TOA_{ad}][H^{+}] - k_{-1}\,[TOAH^{+}_{ad}] \tag{D4}$$

$$R_2 = k_2\,[TOA^{+}_{ad}][Cl^{-}] - k_{-2}\,[TOAHCl_{ad}] \tag{D5}$$

$$R_3 = k_3\,[TOAHCl_{ad}][\overline{TOA}] - k_{-3}\,[TOA_{ad}][\overline{TOAHCl}] \tag{D6}$$

With

$$R_1 = R_2 = R_3 = R \tag{D7}$$

follows

$$R = k_2\,[Cl^{-}]\,\frac{k_1\,[TOA_{ad}][H^{+}] - R}{k_{-1}} - k_{-2}\,\frac{R + k_{-3}\,[TOA_{ad}][\overline{TOAHCl}]}{k_3\,[\overline{TOA}]} \tag{D8}$$

when from equ. (D4) and from equ. (D6) $[TOAH^{+}{}_{ad}]$ and $[TOAHCl_{ad}]$ are respectively substituted in equ. (D5). The superposition of the equilibria R_1 and R_2 gives the overall HCl extraction equilibrium:

$$K_{ex} = K_{EQ1}\,K_{EQ2}\,K_{EQG3} = \frac{k_1}{k_{-1}}\,\frac{k_2}{k_{-2}}\,\frac{k_3}{k_{-3}} \tag{D9}$$

Rearranging equ. (D8) gives:

$$R = \frac{[\overline{TOA}][H^{+}][Cl^{-}] - \dfrac{1}{K_{ex}}[\overline{TOAHCl}]}{\dfrac{1}{k_1}[\overline{TOA}][Cl^{-}] + \dfrac{1}{K_{EQ1}\,k_2}[\overline{TOA}] + \dfrac{1}{K_{EQ1}\,K_{EQ2}\,k_3}}\,[TOA_{ad}] \tag{D10}$$

If the interface is totally adsorbed with the reactive species and all species consume about the same space [Roos (2000)], then

$$[TOA_{ad}]+[TOA^{+}_{ad}]+[TOAHCl_{ad}]=\max \tag{D11}$$

where max is the maximum possible interfacial coverage. $[TOAH^{+}{}_{ad}]$ from equ. (D4) and $[TOAHCl_{ad}]$ from equ. (D6) substituted in equ. (D11) gives:

$$[TOA_{ad}]=\frac{\max+R\left(\dfrac{1}{k_{-1}}-\dfrac{1}{k_3\,[\overline{TOA}]}\right)}{1+K_{EQ1}\,[H^{+}]+\dfrac{[\overline{TOACl}]}{K_{EQ3}\,[\overline{TOA}]}} \tag{D12}$$

and with this $[TOA_{ad}]$ replaced in equ. (D10) gives

$$R=\frac{[\overline{TOA}][H^{+}][Cl^{-}]-\dfrac{1}{K_{ex}}[\overline{TOAHCl}]}{denom}\max \tag{D13}$$

where

$$\begin{aligned}denom=&\frac{1}{k_1}\left\{[\overline{TOA}][Cl^{-}]+\frac{1}{K_{EQ3}}[\overline{TOAHCl}][Cl^{-}]+\frac{K_{EQ1}}{K_{ex}}[\overline{TOAHCl}]\right\}\\&+\frac{1}{k_2}\left\{\frac{1}{K_{EQ1}}[\overline{TOA}]+[\overline{TOA}][H^{+}]+\frac{1}{K_{EQ1}\,K_{EQ3}}[\overline{TOAHCl}]\right\}\\&+\frac{1}{k_3}\left\{\frac{1}{K_{EQ1}\,K_{EQ2}}+\frac{1}{K_{EQ2}}[H^{+}]+[H^{+}][Cl^{-}]\right\}\end{aligned} \tag{D14}$$

When considering the first reaction step (equ. (D1)) to be instantaneous $(k_1\rightarrow\infty)$ this simplifies to:

$$\begin{aligned}denom=&\frac{1}{k_2}\left\{\frac{1}{K_{EQ1}}[\overline{TOA}]+[\overline{TOA}][H^{+}]+\frac{1}{K_{EQ1}\,K_{EQ3}}[\overline{TOAHCl}]\right\}\\=&\frac{1}{k_3}\left\{\frac{1}{K_{EQ1}\,K_{EQ2}}+\frac{1}{K_{EQ2}}[H^{+}]+[H^{+}][Cl^{-}]\right\}\end{aligned} \tag{D15}$$

This formalism can be applied to all solutes which form a 1:1 complex with *TOA*. Additional reactions may additionally occur in the bulk phases and are not considered in the interfacial kinetics term.

A further extension of this model is the co-extraction of two solutes (e.g. *HCl*, *HAc*) and it can be easily extended to further species. For a co-extraction of acetic acid the elementary steps

$$TOAH^+_{ad} + Ac^- \leftrightarrow TOAHAc_{ad} \tag{D16}$$

$$TOAHAc_{ad} + TOA \leftrightarrow TOAHAc + TOA_{ad} \tag{D17}$$

and their reaction rates

$$R_4 = k_4\,[TOAH^+_{ad}][Ac^-] - k_{-4}\,[TOAHAc_{ad}] \tag{D18}$$

$$R_5 = k_5\,[TOAHAc_{ad}][\overline{TOA}] - k_5\,[TOA_{ad}][\overline{TOAHAc}] \tag{D19}$$

have to be considered, too. The overall extraction equilibrium is then:

$$K_{ex,Ac} = K_{EQ1}\,K_{EQ4}\,K_{EQ5} = \frac{k_1}{k_{-1}}\frac{k_4}{k_{-4}}\frac{k_5}{k_{-5}} \tag{D20}$$

For this parallel extraction the following are valid:

$$R_1 = R_2 + R_4 = R_3 + R_5 = R \tag{D21}$$

$$R_2 = R_3 = R_{Cl} \tag{D22}$$

$$R_4 = R_5 = R_{Ac} \tag{D23}$$

$[TOAHCl_{ad}]$ from equ. (D5) in equ. (D6) gives:

$$R_{Cl} = k_3\,[\overline{TOA}]\,\frac{k_2\,[TOAH^+_{ad}][Cl^-] - R_{Cl}}{k_{-2}} - k_{-3}\,[TOA_{ad}][\overline{TOAHCl}] \tag{D24}$$

and

$$R_{Cl} = \frac{a\,k_2\,[TOAH^+_{ad}][Cl^-] - k_{-3}\,[TOA_{ad}][\overline{TOAHCl}]}{1+a} \tag{D25}$$

where

$$a = \frac{k_3}{k_{-2}}[\overline{TOA}] \tag{D26}$$

Analogous with equ. (D17) and (D18) is

$$R_{Ac} = \frac{b\,k_4\,[TOAH^+_{ad}][Ac^-] - k_{-5}\,[TOA_{ad}][\overline{TOAHAc}]}{1+b} \tag{D27}$$

where

$$b = \frac{k_5}{k_{-4}}[\overline{TOA}] \tag{D28}$$

At the interface there are now four species:

$$[TOA_{ad}]+[TOA^{+}_{ad}]+[TOAHCl_{ad}]+[TOAHAc_{ad}]=\max \tag{D29}$$

$[TOAHCl_{ad}]$ from equ. (D5) and $[TOAHAc_{ad}]$ from equ. (D18) together with equ. (D29) give:

$$\begin{aligned}\max = {} & [TOA_{ad}]+[TOAH^{+}_{ad}]+\frac{k_2\,[TOAH^{+}_{ad}][Cl^{-}]-R_{Cl}}{k_{-2}}\\ & +\frac{k_4\,[TOAH^{+}_{ad}][Cl^{-}]-R_{Ac}}{k_{-4}}\end{aligned} \tag{D30}$$

and with equ. (D27) after rearranging:

$$\begin{aligned}\max = {} & [TOA_{ad}]\left\{1+\frac{k_{-5}}{(1+b)\,k_{-4}}[\overline{TOAHAc}]\right\}-\frac{R_{Cl}}{k_{-2}}\\ & +[TOAH^{+}_{ad}]\left\{1+K_{EQ2}\,[Cl^{-}]+\frac{1}{1+b}K_{EQ4}\,[Ac^{-}]\right\}\end{aligned} \tag{D31}$$

Again, the protonation (equ. (D1)) is assumed to be instantaneous:

$$[TOAH^{+}_{ad}]=K_{EQ1}\,[TOA_{ad}][H^{+}] \tag{D32}$$

and with equ. (D31) gives

$$[TOA_{ad}]=\frac{k_{-2}\max+R_{Cl}}{k_{-2}\,c} \tag{D33}$$

where

$$\begin{aligned}c = {} & 1+\frac{k_{-5}}{(1+b)\,k_{-4}}[\overline{TOAHAc}]+K_{EQ1}\,[H^{+}]\\ & +K_{EQ1}\,K_{EQ2}\,[H^{+}][Cl^{-}]+\frac{1}{1+b}K_{EQ1}\,K_{EQ4}\,[H^{+}][Ac^{-}]\end{aligned} \tag{D34}$$

When inserting equ. (D32) and equ. (D33) in equ. (D25), then:

$$R_{Cl}=\frac{k_{-2}\max d}{k_{-2}\,c\,(1+a)-d} \tag{D35}$$

where

$$d=a\,K_{EQ1}\,k_2\,[H^{+}][Cl^{-}]-k_3\,[\overline{TOAHCl}] \tag{D36}$$

With a, b, c and d in equ. (D35) the rate expression for chloride becomes:

$$R_{Cl}=\frac{[\overline{TOA}][H^{+}][Cl^{-}]-\dfrac{1}{K_{ex}}[\overline{TOAHCl}]}{denom_S}\,max \tag{D37}$$

where

$$denom_S = \frac{1}{k_2}\left\{\frac{1}{K_{EQ1}}[\overline{TOA}]+[\overline{TOA}][H^+]+\frac{1}{K_{EQ1}\,K_{EQ3}}[\overline{TOAHCl}]\right\}$$
$$+\frac{1}{k_3}\left\{\frac{1}{K_{EQ1}\,K_{EQ2}}+\frac{1}{K_{EQ2}}[H^+]+[H^+][Cl^-]\right\}$$
$$+\frac{\dfrac{1}{K_{EQ4}\,k_5}+\dfrac{1}{k_4}[\overline{TOA}]}{\dfrac{1}{K_{EQ2}\,k_3}+\dfrac{1}{k_2}[\overline{TOA}]}$$
$$\left\{\frac{1}{K_{EQ1}\,K_{EQ5}\,k_4}[\overline{TOAHAc}]+\frac{1}{k_5}[H^+][Ac^-]\right\} \tag{D38}$$

As can be seen the difference between the chloride extraction in the pure (equs. (D13) and (D15)) and mixed system (equs. (D37) and (D38)) is only in the denominator:

$$denom_S = denom + \frac{\dfrac{1}{K_{EQ4}\,k_5}+\dfrac{1}{k_4}[\overline{TOA}]}{\dfrac{1}{K_{EQ2}\,k_3}+\dfrac{1}{k_2}[\overline{TOA}]}$$
$$\left\{\frac{1}{K_{EQ1}\,K_{EQ5}\,k_4}[\overline{TOAHAc}]+\frac{1}{k_5}[H^+][Ac^-]\right\} \tag{D39}$$

When we apply this procedure also for the second solute, we obtain for the extraction of acetate:

$$R_{Ac} = \frac{[TOA][H^+][Ac^-]-\dfrac{1}{K_{ex,Ac}}[\overline{TOAHAc}]}{denom_{S2}}\max \tag{D40}$$

where

$$denom_{S2} = \frac{1}{k_4}\left\{\frac{1}{K_{EQ1}}[\overline{TOA}] + [\overline{TOA}][H^+] + \frac{1}{K_{EQ1}\,K_{EQ5}}[\overline{TOAHAc}]\right\}$$
$$+\frac{1}{k_5}\left\{\frac{1}{K_{EQ1}\,K_{EQ4}} + \frac{1}{K_{EQ4}}[H^+] + [H^+][Ac^-]\right\}$$
$$+\frac{\dfrac{1}{K_{EQ2}\,k_3} + \dfrac{1}{k_2}[\overline{TOA}]}{\dfrac{1}{K_{EQ4}\,k_5} + \dfrac{1}{k_4}[\overline{TOA}]}$$
$$\left\{\frac{1}{K_{EQ1}\,K_{EQ3}\,k_2}[\overline{TOAHCl}] + \frac{1}{k_3}[H^+][Cl^-]\right\} \tag{D41}$$

The LAP model can be extended in a similar manner to a further co-extraction of solutes, which only will alter the denominator as shown with *HCl* and *HAc* extraction.

Appendix E – Physical and Chemical Properties for *Zn/D2EHPA* and *HAc/TOA*

Table E-1: Physical properties $ZnSO_4$/*D2EHPA*/isododecane (8 mg/L–8 g/L $ZnSO_4$, 0.005–0.2 mol/L *D2EHPA*, 1.5 < PH < 7.0, 298 K)

Aqu. density	$\frac{\text{kg}}{\text{m}^3}$	-	$\rho_{aq} = 156.5 \times [ZnSO_4] + 997.2$
Org. density	$\frac{\text{kg}}{\text{m}^3}$	-	$\rho_{org} = 75.7 \times [D2EHPA] + 745.4$
org. viscosity	$\frac{\text{mm}^2}{\text{s}}$	-	$\nu_{org} = 1.08 \times [D2EHPA] + 1.637$
Interfacial tension	$\frac{\text{mN}}{\text{m}}$	$[Zn] < 0.001$	$\sigma_{small} = 17.23 \times [D2EHPA]^{-0.094}$
Interfacial tension	$\frac{\text{mN}}{\text{m}}$	$0.001 < [Zn] < 0.01$	$\sigma_{medium} = 18.31 \times [D2EHPA]^{-0.088}$
Interfacial tension	$\frac{\text{mN}}{\text{m}}$	$[Zn] > 0.01$	$\sigma_{l\,arg\,e} = 18.61 \times [D2EHPA]^{-0.092}$

Table E-2: Masson parameters $H_2SO_4/ZnSO_4$ (298 K)

Species	M_j [g/mol]	Φ^0_j [cm^3/mol]	S_j [cm^3 mol^{-1} (mol/l)$^{-1/2}$]	$\Phi_{Zn,org}$ [cm^3/mol]
H^+	1.0079	0	0	0.
Zn^{2+}	65.38	–22.27	4.66	–4.2
SO_4^{2-}	96.0636	13.98	8.64	
HSO_4^-	97.0715	37.88	2.18	

Table E-3: Pitzer parameters $H_2SO_4/ZnSO_4$ (298 K)

Interaction	$\beta^{(0)}$	$\beta^{(1)}$	$\beta^{(2)}$	C^{ϕ} or $C^{\phi(0)}$	$C^{\phi(1)}$	α_1	α_2	ω
$Zn^{2+}-SO_4^{2-}$	0.16724	3.49906	−40.5911	0.036746	−12.9451	1.4	12	3.3
$Zn^{2+}-HSO_4^-$	0.56879	*2.61593*	-	−0.046724	-	2.0	-	-
$H^+-SO_4^{2-}$	0.06421	0.225902	-	0.031126	-	2.0	-	-
$H^+-HSO_4^-$	0.22297	0.460016	-	−0.002660	-	2.0	-	-
$Zn^{2+}-H^+$	Θ	0						
$Zn^{2+}-H^+-SO_4^{2-}$	Ψ	0						
$Zn^{2+}-H^+-HSO_4^-$	Ψ	0						
$SO_4^{2-}-HSO_4^-$	Θ	−0.135342						
$Zn^{2+}-SO_4^{2-}-HSO_4$	Ψ	0.0731378						
$H^+-SO_4^{2-}-HSO_4^-$	Ψ	0.0278059						

Table E-4: Hildebrand–Scott parameters *D2EHPA*/diluent

Species	M_i [g/mol]	V_i [cm^3/mol]	δ_i [cal$^{1/2}$cm$^{-3/2}$]
D2EHPA (monomer)	322.43	332.61	8.76
n-heptane	100.203	147.44	7.45
isododecane	170.34	228.422	7.031

System	δ_c	Equilibrium constant
$ZnSO_4$ / *D2EHPA* in n-heptane	9.086 cal$^{1/2}$cm$^{-3/2}$	$10^{-0.9441}$ (mol/L)$^{1/2}$
$ZnSO_4$ / *D2EHPA* in isododecane	9.304 cal$^{1/2}$cm$^{-3/2}$	$10^{-1.1863}$ (mol/L)$^{1/2}$
rhodamine/*D2EHPA* in isododecane	31.2 cal$^{1/2}$cm$^{-3/2}$	$10^{-2.598}$

Table E-5: Diffusion coefficients *Zn*/*D2EHPA* in n-heptane

Binary system	V_1 [cm^3/mol]	M_2 [g/mol]	η_2 [mPa·s]	$D° = Đ$ [m^2/s]
R_2H_2 in isododecane	806.4	170.34	1.2191	$4.25732\cdot10^{-10}$
$ZnR_2(RH)$ in isododecane	1191.1	170.34	1.2191	$3.36898\cdot10^{-10}$
$ZnR_2(RH)$ in R_2H_2	1191.1	644.86	39.7	$2.014\cdot10^{-11}$

Table E-6: Pitzer parameters *NaCl/NaAc*

Species	$\beta^{(0)}$	$\beta^{(1)}$	τ
$Na^+ - Ac^-$ (I)	0.1426	0.3237	–0.00639
$Na^+ - Cl^-$ (I)	0.0765	0.2664	0.00127
$H^+ - Cl^-$ (I)	0.1775	0.2945	0.008
$HAc - HAc$ (II)	0.0608		

I Pitzer & Mayorga (1973)
II Ziegenfuß & Maurer (1994)

Table E-7: Dissociation equilibria and solubility parameters in the *TOA* system

	In toluene		In isododecane	
a:b	$\ln K_{a:b}$ [-]	δ $[cal^{1/2}cm^{-3/2}]$	$\ln K_{a:b}$ [-]	δ $[cal^{1/2}cm^{-3/2}]$
Toluene	-	8.9	-	-
Isododecane	-	-	-	7.0
TOA	-	8.5	-	8.5
HA	2.726	10.1	1.883	10.1
$(HA)_2$	7.479	10.1	6.633	10.1
TOAHA	4.778	9.567	3.768	9.567
$TOA(HA)_2$	-	-	7.845	9.593
$TOA(HA)_3$	14.953	9.610	-	-
$TOA(HA)_4$	-	-	16.750	9.881

Table E-8: Formed species and equilibrium constants of the aqueous phase

Species *i*	pK_i
HAc	4.784
$HCit^{2-}$	6.396
H_2Cit^-	11.157
H_3Cit	14.285

Table E-9: Kinetic constants *TOA–Hac*

	K_1 [kg/mol]	k_2 [kg/(mol s)]	k_{-2} [1/s]	k_3 [kg/(mol s)]	k_{-3} [kg/(mol s)]	σ [-]
Toluene	0.9078×10^{1}	0.4231×10^{3}	0.1269×10^{-1}	0.7147×10^{-1}	0.3607×10^{0}	5.38
Isododecane	0.3884×10^{1}	0.6497×10^{2}	0.8591×10^{-2}	0.1709×10^{4}	0.8566×10^{4}	6.12

Table E-10: LAP model parameters

Parameter	Acetic acid	Hydrochloric acid	Citric acid
max [mmol/m^2]	1.24	1.24	1.24
ln K_{EQ1}	6.2777	6.2777	6.2777
ln k_2	7.6170	6.7462	6.5002
ln K_{EQ2}	2.2930	0.3789	0.2294
ln k_3	4.2538	4.7097	4.6405
ln K_{ex}	11.0240	9.1053	7.8610

Literature

Abou-Nemeh, I., Bart, H.-J. (1998) Microstructures in the System Water/D2EHPA/Span-80/n-Dodecane. Langmuir, 14, 4451–4459

Abrahams, D.S., Prausnitz, J.M. (1975) Statistical Thermodynamics of Liquid Mixtures: A New Expression for the Excess Gibbs Energy of Partly or Completely Miscible Systems. AIChE J., 21, 116–128

Agbe, D., Mendes-Tatsis, M.A. (2000) The Effect of Surfactants on Interfacial Mass Transfer in Binary Liquid–Liquid Systems. Int. J. Heat & Mass Transfer, 43, 1025–1034

Ajawin, L.A., Perez de Ortiz, E.S., Sawistowski, H. (1980) Kinetics of Extraction of Zinc by Di(2ethylhexyl)phosphoric acid in n-Heptane. Proc. ISEC '80 Vol. III, Paper 80–112, University Liege , Liege

Akita, S., Takeuchi, H. (1990) Sorption and Separation of Metals from Aqeous Solution by a Macromolecular Resin Containing Tri-n-octylamine. J. Chem. Eng. Japan, 23

Akita, S., Hirano, K., Ohashi, Y. (1993) Equilibrium Distribution of Palladium(II) between Hydrochloric Acid Solution and a Macromolecular Resin Containing Tri-n-octylamine. Solvent Extraction and Ion Exchange, 11 (5), 797–810

Alonso, A.I., Urtiaga, A.M., Irabien, A., Ortiz, M.I. (1994) Extraction of Cr (VI) with Aliquat 336. In: Hollow Fiber Contactors: Mass Transfer Analysis, Chem. Eng. Sci., 49, 901–909

Ananthapadmanaban, K.P., Goddard, E.D. (1987) Aqueous Biphase Formation in PEG/Inorganic Salt Systems. Langmuir, 3, 25

Aparicio, J., Muhammed, M. (1989) Extraction Kinetics of Zinc from Aqueous Perchlorate Solutions by D2EHPA Dissolved in Isopar-H. Hydrometallurgy, 21, 385–399

Arnold, K.R., Toor, H.L. (1967) Unsteady Diffusion in Ternary Gas Mixtures. AIChE J., 13, 909–914

Ashbrook, A.W., Itzkovitch, I.J., Sowa, W. (1979) In: Proc. ISEC '77, CIM Spec. Vol. 21, Can. Inst. Min. Met., Toronto, 781–791

Asprion, N., Hasse, H., Maurer, G. (1998) Limiting Activity Coefficients in Alcohol Containing Organic Solutions from Head Space Gas Chromatography. J. Chem. Eng. Data, 43, 74–80

Baes, C.F., Moyer, B.A., Case, G.N., Case, F.I. (1990) SXLSQA: A Computer Program for Including both Complex Formation and Activity Effects in the Interpretation of Solvent Extraction Data. Sep. Sci. Technol., 25, 1675–688

Bahmanyar, H., Slater, M.J. (1991) Studies of Break-up in Liquid–Liquid Systems in a Rotating Disc Contactor. Part I: Conditions of no Mass Transfer. Chem. Eng. Technol., 14, 79–89

Bart, H.-J., Draxler, J, Marr, R. (1988) Residence Time Selection in Liquid Membrane Permeation for Copper Recovery. Hydrometallurgy, 19, 351–360

Bart, H.-J., Ramaseder, C., Marr, R. (1993) Phenomenological and Mathematical Description of the Osmotic Influence in the Liquid Membrane Technique. Sep. Sci. Technol., 28 (1 - 3), 929–945

Bart, H.-J., Berger, R., Misek, T., Slater, M.J., Schröter, J., Wachter, B. (1994) Recommended Systems for Liquid Extraction Studies. In: Liquid–Liquid Extraction Equipment, J.C. Godfrey and M.J. Slater (Eds), J. Wiley & Sons, London, 1–43

Bart, H.-J., Jüngling, H., Ramaseder, N., Marr, R. (1995) Water and Solute Solubilization and Transport in Emulsion Liquid Membranes. J. Membr. Sci., 102, 103–112

Bart, H.-J., Rousselle, H.-P. (1999) Microkinetics and Reaction Equilibria in the System $ZnSO_4$/D2EHPA/Isododecane. Hydrometallurgy, 51, 285–298

Bart, H.-J. (2000) Reactive Mass Transfer at Fluid Interfaces. In: Transportmechanism Across Fluid Interphases, DECHEMA Monographs Vol. 136, Wiley-VCH, Weinheim, 297–315

Barton, A.F.M. (1991) Handbook of Solubility Parameters and Other Cohesion Parameters., 2nd Ed., CRC Press, Boca Raton

Basu, R., Prasad, R., Sirkar, K.K. (1990) Non-Dispersive Membrane Solvent Back Extraction of Phenol. AIChE J., 36, 450–460

Basu, R., Sirkar, K.K. (1991) Hollow Fiber Contained Liquid Membrane Separation of Citric Acid. AIChE J., 37, 383–93

Basu, R., Sirkar, K.K. (1992) Citric Acid Extraction with Microporous Hollow Fibers. Solvent Extraction and Ion Exchange, 10 (1), 119–143

Bauer, U., Marr, R., Rückel, W., Siebenhofer, M. (1989) Reactive Extraction of Citric Acid from an Aqueous Fermentation Broth. Ber. Bunsenges. Phys. Chem., 93, 980–984

Bird, R.B., Steward, W.E., Lightfoot, E.N. (1960) Transport Phenomena. John Wiley & Sons, Inc., New York

Blaß, E., Liebl, T., Häberl, M. (1997) Extraktion – ein historischer Rückblick. Chem.-Ing.-Techn., 69, 431–437

Blumberg, R., Gai, J.F. (1979) “Strong” Diluent Effects. In: Proc. ISEC '77, 9 - 17 CIM Spec. Vol. 21, Can. Inst. Min., Toronto

Boussinesq, M.J. (1905) Sur l’existence d’une viscosité superficielle dans la mine couche de transition separant un liquide d’une autre fluide contigu. Ann. Chim. Phys., 29, 349–357

Boussinesq, M.J. (1913) Hydrodynamique – Vitesse de la chute lente, devenue uniforme, d’une goutte liquide sphérique, dans un fluide visqueux de poids spécifique moindre. Comptes Rendus Académie des Sciences, 1124–1130

Carr, P.W. (1993) Solvatochromism LSER and Chromatography. Microchemical J., 48, 3–28

Cavers, S.D. (1983) Nonmechanically Agitated Columns. In: Handbook of Solvent Extraction, Chap. 10, T.C. Lo et al. (Ed), John Wiley & Sons, New York

Cauwenberg, V. (1995) Hydrodynamics and Physico-Chemical Aspects of Liquid-Liquid Extraction. PhD thesis, Leuven University, Leuven

Chapman, S., Cowling, T.G. (1952) The Mathematical Theory of Non-Uniform Gases. Cambridge University Press, Cambridge

Chen, S.-H., Evans, D.F., Davis, H.T. (1983) Tracer Diffusion in Methanol, 1-Butanol and 1-Octanol from 298 to 433 K. AIChE J., 29, 640–645

Cianetti, C., Danesi, P.R. (1983) Kinetics and Mechanism of the Interfacial Mass Transfer of Zn(II) Co(II) and Ni(II) in the System: D2EHPA n-Dodecane-KNO_3-Water. Solvent Extraction and Ion Exchange, 1 (1), 9–26

Clausen, I., Kampt, A., Kolbe, B., Arlt, W. (1999) A priori Vorhersage thermodynamischer Daten für thermische Trennverfahren. "Thermische Zerlegung von Gas- und Flüssigkeitsgemischen", Münster

Coelhoso, I.M., Silvestre, P., Viegas, R.M.C., Crespo, J.P.S.G., Carrondo, M.J.T (1997) Membrane-Based Solvent Extraction and Stripping of Lactate in Hollow-Fiber Contactors. J. Membr. Sci., 134, 19–32

Cohen, M.H., Turnbull, D. (1959) Molecular Transport in Liquids and Glasses. J. Phys. Chem., 31, 1164–1169

Coleman, C.F., Leuze, R.E. (1978) Some Milestone Solvent Extraction Processes at the Oak Ridge National Laboratory. J. Tennessee Acad. Sci., 53 (3), 102–107

Cooney, D.O., Jin, C.L. (1985) Solvent Extraction of Phenol from Aqueous Solutions in a Hollow Fiber Device. Chem. Eng. Comm., 37, 173–191

Corsi, C., Gnagnarelli, G., Slater, M.J., Veglio, F. (1998) A Study of the Kinetics of Zinc Stripping for the System Zn/H_2SO_4/D2EHPA/n-Heptane in a Hancil Constant Interface Cell and a Rotating Disc Contactor. Hydrometallurgy, 50, 125–141

Cortina, J.L., Miralles, N., Sastre, A., Aguilar, M., Profumo, A., Pesavento, M. (1992) Solvent Impregnated Resins Containing Cyanex 272. Preparation and Application to the Extraction and Separation of Divalent Metals. Reactive Polymers, 18 (1), 67–75

Cortina, J.L., Miralles, N., Sastre, A., Aguilar, M., Profumo, A., Pesavento, M. (1993a) Solvent Impregnated Resins Containing Di-(2,4,4-trimethylpentyl)phosphinic acid. I. Comparative Study of Di-(2,4,4-trimethylpentyl)phosphinic acid Adsorbed into Amberlite XAD-2 and Dissolved in Toluene. Reactive Polymers, 21 (1–2), 89–10

Cortina, J.L., Miralles, N., Sastre, A., Aguilar, M., Profumo, A., Pesavento, M. (1993b) Solvent Impregnated Resins Containing Di-(2,4,4-trimethylpentyl)phosphinic acid. II. Study of the Distribution Equilibria of Zinc(II), Copper(II) and Cadmium(II). Reactive Polymers, 21 (1–2), 103–116

Cortina, J.L., Miralles, N., Aguilar, M. (1994a) Solvent Impregnated Resin Containing Di-(2-ethylhexyl)phosphoric acid. II. Studies of the Distribution Equilibria of Zn, Cu and Cd. Solvent Extraction and Ion Exchange, 12, 371–391

Cortina, J.L., Miralles, N., Aguilar, M. (1994b) Extraction Studies of Zn, Cu, Cd with Impregnated and Levextrel Resins Containing Di-(2-ethylhexyl)phosphoric acid. Hydrometallurgy, 36, 131–142

Cortina, J.L., Miralles, N., Aguilar, M., Warshawsky (1995a) Solid–Liquid Distribution Studies of Divalent Metals from Nitrate Media Using Impregnated Resins Containing a Bifunctional Organophosphorous Extractant (o-Methyl-dihexyl-phosphine-oxide-o'-hexyl-2-ethyl phos-phoric acid). Reactive Functional Polymers, 27 (1), 61–73

Cortina, J.L., Miralles, N., Aguilar, M. (1995b) Solid–Liquid Extraction Studies of Zn (II), Cu (II) and Cd (II) from Chloride Media with Impregnated Resins Containing Mixtures of Organophosphorus Compounds Immobilized onto Amberlite XAD2. Hydrometallurgy, 37, 301–322

Cortina, J.L., Miralles, N., Aguilar, M. (1996) Distribution Studies of Zn (II), Cu (II) and Cd (II) with Levextrel Resins Containing Di-(2,4,4-trimethylpentyl)phosphinic acid (Lewatit TP807'84), Hydrometallurgy, 37, 301–322

Costello, M.J., Fane, A.G., Hogan, P.A., Schofield, R.W. (1993) The Effect of Shell Side Hydrodynamics on the Performance of Axial Flow Hollow Fiber Modules. J. Membr. Sci., 80, 1–11

Coulaloglou, C.A., Tavlarides, L.L. (1976) Drop Size Distributions and Coalescence Frequencies of Liquid-Liquid Dispersions in Flow Vessels. AIChE J., 22, 289–297

Cox, M., Flett, D.S. (1983) Metal Extractant Chemistry. In: Handbook of Solvent Extraction, T.C. Lo et al. (Ed), J. Wiley & Sons, New York, 53–90
Cox, M. (1992) Solvent Extraction in Hydrometallurgy. In: Principles and Practice of Solvent Extraction. J. Rydberg (Ed), Marcel Dekker, New York, 357–412
Cullinan, H.T., Cusick, M.R. (1967) Predictive Theory for Multicomponent Diffusion Coefficients. Ind. Eng. Chem. Fundam., 6, 72–77
Cummings, P.T., Evans, D.J. (1992) Nonequilibrium Molecular Dynamics Approaches to Transport Properties and Non Newtonian Fluid Rheology. Ind. Eng. Chem. Res., 31, 1237–1252
Cussler, E.L. (1971) Membranes which Pump. AIChE J., 17, 1300–1303
Cussler, E.L. (1976) Multicomponent Diffusion. Elsevier, Amsterdam
Cussler, E.L. (1997) Diffusion–Mass Transfer in Fluid Systems. 2nd Ed., Cambridge Univ. Press, Cambridge
Czapla, C. (2000) Zum Einfluss von Grenzflächenladungen auf den Reaktionsmechanismus organischer Säuren bei der Reaktivextraktion. Shaker Verlag, Aachen
Czapla, C., Bart, H.-J. (2000a) Characterisation and Modeling of the Extraction Kinetics of Organic Acids Considering Boundary Layer Charge Effects. Chem. Eng. Technol., in press
Czapla, C., Bart, H.-J. (2000b) The Influence of the Surface Potential on Interfacial Kinetics. Ind. Eng. Chem. Res., submitted
Czapla, C., Bart, H.-J., Jesberger, M. (2000) Mass Transfer and Zeta Potential in Reactive Extraction. Sep. Sci. Technol., 35 (9), 1423–1438
Dahuron, L., Cussler, E.L. (1988) Protein Extractions with Hollow Fibers. AIChE J., 34 (1), 130–136
Daiminger, U., Nitsch, W., Plucinski, P., Geist, A. (1996) The Efficiency of Hollow Fiber Modules for Nondispersive Chemical Extraction. Ind. Eng. Chem. Res., 35, 184
Danckwerts, P.V. (1951) Significance of Liquid Film Coefficients in Gas Absorption. Ind. Eng. Chem., 43, 1460–1467
Danesi, P.R., Chiarizia, R. Muhammed, M. (1978) Mass Transfer in Liquid Anion Exchange Processes I: Kinetic of the Two-Phase Acid-Base Reaction in the System Trilaurylamine–Toluene–HCl–Water. J. Inorg. Nucl. Chem., 40, 1581–1589
Danesi, P.R., Chiarizia, R. (1980) The Kinetics of Metal Solvent Extraction. CRC Crit. Review Anal. Chem., 10, 1–126
Danesi, P.R., Vandegrift, G.F. (1981) Activity Coefficients of Bis(2-ethylhexyl)phosphoric acid in n-Dodecane. Inorg. Nucl. Chem. Letters, 17, 109–115
Danesi, P.R., Reichley-Yinger, L., Rickert, P.G. (1987) Life Time of Supported Liquid Membranes: The Influence of Interfacial Properties Chemical Composition and Water Transport on the Long Term Stability of the Membranes. J. Membr. Sci., 31, 117–145
De Belval, S., Breton, le, B., Huddleston, J.G., Lyddiatt, A. (1998) The Influence of Temperature upon Protein Partitioning in PEG-Salt Aqueous Two-Phase Systems Close to the Critical Point with some Observations Relevant to the Partitioning of Particles. J. Chromatography, 13, 19, 711
Debye, P., Hückel, E. (1923) The Theory of Electrolytes. I. Lowering of Freezing Point and Related Phenomena. Physik. Z., 24 (9), 185–206
D'Elia, N.A., Dahuron, L., Cussler, E.L. (1986) Liquid–Liquid Extractions with Microporous Hollow Fibers. J. Membr. Sci., 29, 309–319
De Haan, A.B.P., Bartels, P.V., Graauw, de, J. (1989) Extraction of Metal Ions from Waste Water Modeling of the Mass Transfer in a Supported-Liquid Membrane Process. J. Membr. Sci., 45, 281–297

Diamond, R.M. (1967) In: Solvent Extraction Chemistry, D. Dyrrsen et al. (Ed), North Holland Pupl. Comp., Amsterdam, 185–194
Ding, H.B., Carr, P.W., Cussler, E.L. (1992) Racemic Leucine Separation by Hollow-Fiber Extraction. AIChE J., 38, (10), 1493–1498
Doug, D.C., Winnik, M.A. (1984) The Py Scale of Solvent Polarities. Can. J. Chem., 62, 2560
Edwards, T.J., Maurer, G., Newman, J., Prausnitz, J.M. (1978) Vapor Liquid Equilibria in Multicomponent Aqueous Solutions of Volatile Weak Electrolytes. AIChE J., 24, 966–976
Edwards, D.A., Brenner, H., Wasan, D.T. (1991) Interfacial Transport Process and Rheology. Butterworth-Heinemann, Boston
Egner, K., Gaube, J., Pfennig, A. (1997) GEQUAC–An Excess Gibbs Energy Model for Simultaneous Description of Associating and Non-Associating Liquid Mixtures. Ber. Bunsenges. Phys. Chem., 101, 2, 209–218
Einstein, A. (1905) Über die von der molekularkinetischen Theorie der Wärme geforderte Bewegung von in ruhenden Flüssigkeiten suspendierten Teilchen. Ann. Physik, 17, 549–560
Eiteman, M.A., Gainer, J.L. (1992) A Correlation for Predicting Partition Coefficients in Aqueous Two-Phase Systems. Sep. Sci. and Technology, 27, 313
Enskog, D., Sven, K. (1922) Vetenskapsaked Hand., 63, 4
Escalante, H., Ortiz, I., Irabien, A. (1996) Concentration of L-Phe by Non-Dispersive Extraction in Hollow Fiber Modules. In: Value Adding Through Solvent Extraction, D.C. Shallcross, R. Paimin, H. Prvcic (Eds), Vol. II., The University of Melbourne, 1493–1498
Eyring, H., Hirschfelder, J. (1937) The Theory of the Liquid State. Phys. Chem., 41, 249–257
Fei, W., Bart, H.-J. (1998) Prediction of Diffusivities in Liquids. Chem. Eng. Technol., 21, 8, 659–665
Fei, W., Bart, H.-J. (2000) Predicting Diffusivities in Liquids by the Group Contribution Method. Chem. Eng. and Processing, submitted
Feron, P.H.M., Jansen, A.E., Klaasen, R., Ter Meulen, B.P., Akkerhuis, J.J., Maanen, van, H.C.H.J. (1993) Development of Membrane Based Adsorption and Extraction Processes and Equipment. TNO Sci. Report, Apeldoorn NL
Fick, A. (1855) Über Diffusion. Poggendorff's Annalen der Physik (94), 59
Fish, C., McEachren, T., Hassett, J. (1992) Hollow Fiber Membrane System for the Continuous Extraction and Concentration of Organic Compounds from Water. ACS Sym. Series, 58, 155
Fogler, S.H. (1992) Elements of Chemical Reaction Engineering. Prentice Hall, New Jersey
Fourier, J.B.J. (1822) Théorie Analytique de la Chaleur, Paris
Fredenslund, A., Jones, R.L., Prausnitz, J.M. (1975) Group-Contribution Estimation of Activity Coefficients in Nonideal Liquid Mixtures. AIChE J., 21 (6), 1086–1099
Fredenslund, A., Gmehling, J., Rasmussen, P. (1977) Vapor Liquid Equilibria Using UNIFAC – A Group Contribution Method. Elsevier, Amsterdam
Frumkin, A.N., Levich, V.G. (1947) Effect of Surface-Active Substances on Movements at the Boundaries of Liquid Phases. Zhur. Fiz. Khim. 21, 1183–1204
Gabelman, A., Hwang, S.-T. (1999) Hollow Fiber Membrane Contactors. J. Membr. Sci., 159, 61–106
Gayler, R., Roberts, N.W., Pratt, H.R.C. (1953) Liquid-Liquid Extraction. A Further Study of Hold-Up in Packed Columns. Trans. Am. Inst. Chem. Eng., 31, 57–58

Gloe, K., Chartoux, C., Rambusch, T., Ahnis, F., Ibach, S., Vögtle, F. (1999) Massgeschneiderte aromatische Käfigverbindungen – Modelle für die selektive Extraktion von Metallionen. DVCV-Extraktion, 5./6.5.1999, Münster

Gmehling, J., Li, J., Schiller, A. (1993) A Modified UNIFAC Model. 2. Present Parameter Matrix and Results for Different Thermodynamic Properties. Ind. Eng. Chem. Res., 32, 178

Gmehling, J., Onken, U., Arlt, W., Grenzheuser, P., Kolbe, B., Weidlich, U., Rarey-Nies, J. (1977–1999) Vapor–Liquid Equilibrium Data Collection. Chemistry Data Series, Parts 1–21, DECHEMA, Frankfurt

Godfrey, J.C., Slater, M.J. (Eds) (1994) Liquid–Liquid Extraction Equipment, J. Wiley & Sons, Chichester

Goklen, K.E., Hatton, T.A. (1986) Liquid Extraction of Proteins Using Reverse Micelles. In: Proc. ISEC '86, Vol. 3, 587–595, Ed. DECHEMA, Frankfurt

Grimm, R., Kolarik, Z. (1974a) Acidic Organophosphorus Extractants Complexes of Some Bivalent Transition Metals with D2EHPA in Highly Loaded Organic Phases. Proc. ISEC '74, Vol. I: 767–780, Soc. Chem. Ind., London

Grimm, R., Kolarik, Z. (1974b) Acidic Organophosphorous Extractants XIX: Extraction of Cn(II) Ni(II), Zn(II) and Ld(II) by D2EHPA. J. Inorg. Nucl. Chem., 36, 189

Großmann, C., Tintinger, R., Zhu, J., Maurer, G. (1995) Aqueous Two-Phase Systems of Poly(ethylene glycol) and Dextran: Experimental Results and Modeling of Thermodynamic Porperties. Fluid Phase Equilibria, 106, 111-138

Großmann, C., Tintinger, R., Zhu, J., Maurer, G. (1995) Aqueous Two-Phase Systems of Poly(ethylene glycol) and Di-Potassium Hydrogen Phosphate with and without Partitioning Biomolecules – Experimental Results and Modeling of Thermodynamic Properties. Ber. Bunsenges. Phys. Chemie, 99, 700–712

Guggenheim, E.A. (1935) Specific Thermodynamic Properties of Aqueous Solutions of Strong Electrolytes. Phil. Mag., 19, 588–643

Guggenheim, E.A. (1952) Mixtures. Clarendon Press, Oxford

Guggenheim, E.A., Stokes, R.H. (1969) Equilibrium Properties of Aqueous Solutions of Single Strong Electrolytes. Pergamon Press, Oxford

Guo, X.M., Fei, W.Y., Wang, J.D. (1997) Prediction of Mutual Diffusion Coefficients of Aromatics by Group Contribution Method. Proceedings of International Symposium on Liquid–Liquid Two Phase Flow and Transport Phenomena, 41 - 43, Turkey

Hait, M.J., Liotta, C.L., Eckert, C.A., Bergman, D.L., Karachewski, A.M., Dallas, A.J., Eikens, D.I., Li, J.J., Carr, P.W., Poe, R.B., Rutan, S.C. (1993) SPACE Predictor for Infinite Dilution Activity Coefficients. Ind. Eng. Chem. Res., 3, 2905–2914

Hancil, V., Slater, M.J., Yu, W. (1990) On the Possible Use of Di(2-ethylhexyl)phosporic acid as a Recommended System for Liquid–Liquid Extraction: The Effect of Impurities on Kinetics. Hydrometallurgy, 25, 375–386

Handlos, A.E., Baron, T. (1957) Mass and Heat Transfer from Drops in Liquid–Liquid Extraction. AIChE J., 3 (1), 127–135

Harned, H.S., Owen, B.B. (1958) The Physical Chemistry of Electrolyte Solutions. Reinhold Publishing, New York

Hasse, H. (1996) Anwendung der Spektroskopie in thermodynamischen Untersuchungen fluider Mischungen. VDI-Fortschrittsbericht Nr. 458, Reihe 3, VDI-Verlag, Düsseldorf

Hayduk, W., Minhas, B.S. (1982) Correlation for Prediction of Molecular Diffusivities in Liquids. Can. J. Chem. Eng., 60, 295

Henschke, M. (1995) Dimensionierung liegender flüssig-flüssig Abscheider anhand diskontinuierlicher Absetzversuche. VDI-Verlag, Düsseldorf

Henschke, M., Pfennig, A. (1999) Mass Transfer Enhancement in Single-Drop Extraction Experiments. AIChE J., 45 (10), 2079–2086

Higbie, R. (1935) The Rate of Absorption of a Pure Gas into a Still Liquid During Short Periods of Exposure. Trans. Am. Inst. Chem. Eng., 31, 365–383

Hildebrand, J.M., Scott, R.L. (1950) The Solubility of Nonelectrolytes, 3rd Ed., Reinhold, New York

Hinze, W.L., Pramauro, E.A. (1993) A Critical Review of Surfactant Mediated Phase Separations (Cloud Point Extraction): Theory and Applications. CRC Crit. Rev. Anal. Chem., 24, 133

Hirschfelder, J., Curtis, C.F., Bird, R.B. (1954) Molecular Theory of Gases and Liquids. Wiley & Sons, New York

Ho, W.S.W., Li, N.N. (1992) Theory. In: Membrane Handbook. W.S.W. Ho, K.K. Sirkar (Eds), Chap. 37, Van Nostrand Reinhold, New York, 611–655

Högtfeld, E. (1976) Diluents in the Solvent Extraction of Metals. Diluent Testing and its Problems. Chem. Ind., 170–174

Holmyard, E.J. (1957) Alchemy. Pelican Books, A 348, London

Horst, S., Fischer, K., Gmehling, J. (2000) PSRK Group Contribution Equation of State: Revisions and Extensions III. Fluid Phase Equilibria, 167, 173

Horvath, A.L. (1985) Handbook of Aqueous Electrolyte Solutions. Ellis Horwood, London

Howell, W.J., Karachewski, A.M., Stephenson, K.M., Eckert, C.A. (1989) An Improved MOSCED Equation for the Prediction and Application of Infinite Dilution Activity Coefficients. Fluid Phase Equilibria, 52, 151–160

Huang, T.C., Juang, R.S. (1986) Kinetics and Mechanism of Zinc Extraction from Sulfate Medium with D2EHPA. J. Chem. Eng. Japan, 19 (5), 379 -386

Huddleston, J.G., Veide, A., Köhler, K., Flanagan, J., Enfors, S.-O., Lyddiatt, A. (1991) The Molecular Basis of Partitioning in Aqueous Two-Phase Systems. Trends Biotechnol., 9, 381

Huddleston, J.G., Willauer, H.D., Swatloski, R.P., Visser, A.E., Rogers, R.D (1998) Room Temperature Ionic Liquids as Novel Media for 'Clean' Liquid–Liquid Extractions. Chem. Commun., 1765–1766

Huddleston, J.G., Willauer, H.D., Griffin, S.T., Roger, R.D. (1999) Aqueous Polymeric Solutions as Environmentally Benign Liquid/Liquid Extraction Media. Ind. Eng. Res., 38, 2523–2539

Hulburt, H.M., Katz, S. (1965) Some Problems in Particle Technology. Chem. Eng. Sci., 19, 555–574

Hunter, R.J. (1981) Zeta Potential in Colloid Science. Oxford University Press Inc., New York

Hunter, R.J. (1993) Introduction to Modern Colloid Science. Oxford University Press Inc., New York

Humphrey, J.L., Rocha, J.A., Fair, J.R. (1984) The Essentials of Extraction. Chem. Eng. Sep., 17, 76–95

Hurter, P.N., Hatton, T.A. (1992) Solubilization of Polycyclic Aromatic Hydrocarbons by Poly(ethylene oxide-propylene oxide) Block Copolymer Micelles: Effects of Polymer Structure. Langmuir, 8, 1291

Ishii, H., Junichiro, M., Wantanabe, H. (1977) Extraction of Nickel with 1-(2-Thiazolylazo)-2-naphtol into a Nonionic Surfactant as the Solvent. Bunseki Kagaku, 26, 252

Johansson, H.-O., Karlström, G., Mattiasson, B., Tjerneld, F. (1995) Effects of Hydrophobicity and Counter Ions on the Partitioning of Amino Acids in Thermoseparating Ucon–Water Two-Phase System. Bioseparation, 5, 269

Juang, R.S., Su, J.Y. (1992a) Separation of Zinc and Copper from Aqeous Sulfate Solutions Using Bis(2-ethylhexyl)phosphoric acid-Impregnated Macroporous Resins. Ind. Eng. Chem. Res., 31 (12), 2779–2783

Juang, R.S., Su, J.Y. (1992b) Sorption of Copper and Zinc from Aqeous Sulfate Solutions with Bis(2-ethylhexyl)phosphoric acid-Impregnated Macroporous Resins. Ind. Eng. Chem. Res., 31 (12), 2774–2749

Juang, R.S., Chang, M.L. (1995) Distribution Equilibrium of Citric Acid Between Aqueous Solutions on Tri-n-octylamine-Impregnated Macroporous Resins. Ind. Eng. Chem. Res., 34, 1294–1301

Juang, R.S., Lin, H.C. (1995a) Metal Sorption with Extractant-Impregnated Resins. 1. Particle Diffusion Kinetics. J. Chem. Tech. Biotechnol., 62, 132–140

Juang, R.S., Lin, H.C. (1995b) Metal Sorption with Extractant-Impregnated Resins. 2. Chemical Reaction and Particle Diffusion Kinetics. J. Chem. Tech. Biotechnol., 62, 141–147

Juang, R.S., Chang, H.L. (1995c) Distribution Equilibrium of Citric Acid Between Aqeous Solutions and Tri-n-octylamine-Impregnated Macroporous Resins. Ind. Eng. Chem. Res., 34, 1294–1301

Juang, R.S., Lee, S.H. (1996a) Column Separation of Divalent Metals from Sulfate Solutions Using Impregnated Resins Containing Di-(2-ethylhexyl)phosphoric acid. Reactive Polymers, 29, 175–183

Juang, R.S., Lee, S.H., (1996b) Column Sorption of Divalent Metals from Sulfate Solutions by Impregnated Macroporous Resins. J. Chem. Tech. Biotechnol., 66, 153–159

Juang, R.S., Chou, T.C. (1996c) Sorption Kinetics of Citric Acid from Aqueous Solutions by Macroporous Resins Containing a Tertiary Amine. J. Chem. Eng. Japan, 29, Nr. 1, 146–151

Juang, R.S., Chen, M.L. (1997) Application of the Elovich Equation to the Kinetics of Metal Sorption with Solvent-Impregnated Resins. Ind. Eng. Chem. Res., 36 (3), 813–820

Jungfleisch, E., Berthelot, M. (1872) Sur les lois qui président auf partage d'un corps entre deux dissolvants. Annales de Chimie, XXVI, 396–407

Kahleweit, M., Strey, R., Busse, G. (1990) Microemulsions: A Qualitative Thermodynamic Approach. J. Phys. Chem., 94, 3881–3894

Kamenski, D.I., Dimitrov, S.D. (1993) Parameter Estimation in Differential Equations by Application of Rational Functions. Computers Chem. Engng., 17, 643–651

Kamlet, M.J., About, J.L.M., Taft, R.W. (1981) An Examination of Linear Solvation Energy Relationship. Prog. Phys. Org. Chem., 13, 485–630

Kamlet, M.J., About, J.L.M., Abraham, M.H., Taft, R.W. (1983) LSER 23 A Comprehensive Collection of the Solvatochromic Parameters π^*, α, ß and Some Methods for Simplifying the Generalized Solvatochromic Equation. J. Org. Chem., 48, 2877–2887

Kamlet, M.J., Doherty, R.M., Abraham, M.H., Marcus, Y., Taft., R.W. (1988) LSER 46 An Improved Equation for Correlating and Prediction of Octanol/Water Partition Coefficients of Organic Nonelectrolytes (Including Strong Hydrogen Bond Donor Solutes. J. Phys. Chem., 92, 5244–5255

Karachewski, A.M., Howell, W.J., Eckert, C.A. (1991) Development of the AVEC Model for Associating Mixtures Using NMR Spectroscopy. Ind. Eng. Chem. Res., 37, 65–73

Kathios, D.J., Jarvinen, G.D., Yarbro, S.L., Smith, B.F. (1994) A Preliminary Evaluation of Microporous Hollow Fiber Membrane Modules for the Liquid–Liquid Extraction of Actinides. J. Membr. Sci., 97, 251–261

Kehiaian, H.V., Grolier, J.B., Benson, G.C. (1978) Thermodynamics of Organic Mixtures. A Generalized Quasichemical Theory in Terms of Group Surface Interactions. J. Chimie Physique, 75, 1031–1048

Keil, F. (1999) Diffusion und Chemische Reaktionen in der Gas/Feststoff-Katalyse. Springer Verlag, Heidelberg

Khalifa, S.M., Shehata, F., Aly, H.F. (1992) Diluent Influence on the Extraction of Divalent Co by Thenonyltrifluoracetone, Dibenzo-18-crown-6 and their Mixtures from Perchlorate Aqueous Medium. In: Solvent Extraction 1990, T. Sekine (Ed), Elsevier, Amsterdam, 219–224

Kim, B.M. (1984) Membrane-based Solvent Extraction for Selective Removal and Recovery of Metals. J. Membr. Sci., 21, 15–19

Kirsch, T., Ziegenfuß, H., Maurer, G. (1997) Distribution of Citric, Acetic and Oxalic Acid between Water and Organic Solutions of Tri-n-octylamine. Fluid Phase Equilibria, 129, 235–266

Kitazaki, H., Ishimaru, M., Inoue, K., Yoshida, K., Nakamura, S. (1996) Separation and Recovery of Flavonoids by Means of Solvent Extraction and Adsorption on Solvent-Impregnated Resin. Value Adding Through Solvent Extraction, J.E. Shallcross, P. Paimin, L.M. Prvcic (Eds), 1667–1673, University of Melbourne, Melbourne

Klaasen, R., Jansen, A.E. (1998) Pertraction – eine neue Membran-Extraktions-Technologie zur Beseitigung hydrophober organischer Komponenten aus Wasser. DECHEMA-Jahrestagung, Wiesbaden, Band II, 246, DECHEMA , Frankfurt

Klampt, A., Jonas, V., Bürger, T., Lohrenz, J.C.W. (1998) Refinement and Parameterization of COSMO-RS. J. Phys. Chem. A, 102, 5074–5085

Klocker, H. (1996) Multikomponentenstoffaustausch bei der Reaktiv-extraktion im System Zinksulfat/Di(2-ethylhexyl)-phopshorsäure. PhD thesis, Techn. University Graz, Graz

Klocker, H., Bart, H.-J., Marr, R., Müller, H. (1997) Mass Transfer Based on Chemical Potential Theory: $ZnSO_4/H_2SO_4$/D2EHPA. AIChE J., 43, 10, 2479–2487

Kohler, F. (1972) The Liquid State. VCH, Weinheim

Kojima, K., Tochigi, K. (1979) Prediction of Vapor–Liquid Equilibria by the ASOG Method. Elsevier, Tokyo

Kolarik, Z., Grimm, R. (1976) Acidic Organophosphorous Extractants XXIV: The Polymerization Behaviour of Cn(II) Cd(II), Zn(II) and Co(II) Complexes of D2EHPA in Fully Loaded Organic Phases. J. Inorg. Nucl. Chem., 38, 1721–1727

Kolarik, Z. (1982) Critical Evaluation of some Equilibrium Constants Involving Acidic Organophosphorous Extractants. Pure & Appl. Chem., 54 (12), 2593–2674

Koncar, M., Bart, H.-J., Marr, R. (1988) Extraction of Zn by D2EHPA – Influence of Activity and High Loading. Proc. ISEC '88, Vol. III, 175–178, Nauka, Moscow

Kordosky, G.A. (1973) The Chemistry of Metals Recovery Using LIX-Reagents. General Mills, Mines Branch, Canada

Krishna, R. (1977) A Generalized Film Model for Mass Transfer in Non-Ideal Fluid Mixtures. Chem. Eng. Sci., 32, 659–667

Kroebel, R., Meyer, A. (1971)West German Pat. Appl. 2162951 (18-12-1971)

Kronberger, T., Ortner, A., Zulehner, W., Bart, H.-J. (1995) Numerical Simulation of Extraction Columns Using a Drop Population Model. Computers Chem. Engng., 19, 639–644

Kronig, R., Brink, J.C. (1950) On the Theory of Extraction from Falling Droplets. Appl. Sci. Res., A2, 142–154

Kumar, A., Hartland, S. (1994) Empirical Prediction of Operating Variables. In: Liquid–Liquid Extraction Equipment, J.C. Godfrey, M.J. Slater (Eds), John Wiley & Sons, Chichester, 141–226

Laird, W.G. (1919) U.S. Pat. 1.320.396

Larsen, P. (1986) Prediction of Phase Equilibria and Heat Effects of Mixing with a Modified UNIFAC-Model. PhD thesis, Lyngby

Lebens, P.J.M., Keurentjes, J.T.F. (1996) Temperature Induced Solubilization of Hydrocarbons in Aqueous Block Copolymer Solutions. Ind. Eng. Chem. Res., 35, 3415

Leung, R., Hon, J.J., Shah, D.O. (1988) Microemulsions: Formation Structure Properties and Novel Applications. In: Surfactants in Chemical/Process Engineering, D.T. Wasan, M.E. Ginn, D.O. Shah (Eds), Surfactant Science Series, Vol. 28, 315–367, Marcel Dekker, New York

Lewis, W.K., Whitman, W.G. (1924) Principles of Gas Adsorption. Ind. Eng. Chem., 16 (12), 1215–1220

Li, J., Zhang, Y., Carr, P.W. (1993) Development of GC Scale of Solute Hydrogen Bond Acceptor Basicity and Characterization of Some Hydrogen Bond Donor Phases by Use of LSER. Anal. Chem., 65, 1969–1979

Li, J., Polka, H.-M., Gmehling, J. (1994) A g^E Model for Single and Mixed Solvent Electrolyte Systems, Part I. Fluid Phase Equilibria, 94, 89–114

Liem, D.H. (1972) Studies of the Complex Formation Between Di-2-ethylhexyl-phosphate (HDEHP) and Tributylphosphate (TBP) or Trioctylamine (TOA) in Toluene. Acta Chem. Scand., 26, 191–204

Lin, K.L., Osseo-Asare, K. (1986) Interfacial Charge and Mass Transfer in the Liquid–Liquid Extraction of Metals. In: Recent Developments in Separation Science, N.N. Li, J.M. Carlo (Eds), 55–74, CRC Press, Cleveland

Lo, T.C., Baird, M.H.I., Hanson, C. (Eds) (1983) Handbook of Solvent Extraction, J. Wiley & Sons, New York

Lochiel, A.C. (1965) The Influence of Surfactants on Mass Transfer Around Spheres. Can. J. Chem. Eng., 43, 40–44

Lopez, J.L., Matson, S.L. (1997) A Multiphase/Extractive Enzyme Membrane Reactor for Production of Diltiazem Chiral Intermediate. J. Membr. Sci., 125, 189–211

Lorbach, D., Bart, H.-J., Marr, R. (1986) Stoffübergang in der Flüssig-Membran-Permeation. Chem.-Ing.-Techn., 58, 156–157, and VDI-Berichte 607, 1765—1799 and Microfiche MS 1452/86

Luehrs, D.C., Godbole, K.A. (1994) LSER of the Solubility of Very Polar Gases. J. Solv. Chem., 23, 1147–1160

Lyklema, J., De Coninck, J., Rovillard, S. (1998) Electrokinetics. The Properties of the Stagnant Layer Unraveled. Langmuir, 14 (20), 5659–5663

Lyklema, J. (1999) Elektrische Doppelschichten: Elektrostatik und Elektrodynamik. Chem.-Ing.-Techn., 71, 1364–1369

Mack, C. (2000) Untersuchungen zum Stofftransport und Stoffaustauschverhalten bei der Extraktion von Chrom (III) und Zink mittels D2EHPA. PhD Thesis, Techn. University Darmstadt, Darmstadt

MacInnis, M.B., Kim, T.K. (1983) Commercial Processes for Tungsten and Molybdenum. In: Handbook of Solvent Extraction. T.C. Lo, M.H.I. Baird, C. Hanson (Eds), J. Wiley & Sons, New York, 673–688

Magnussen, T., Rasmussen, P., Fredenslund, A. (1981) UNIFAC Parameter Table for Prediction of Liquid–Liquid Equilibria. Ind. Eng. Chem. Proc. Des. Dev., 20, 331–339

Marangoni, C.G.M. (1871) Über die Ausbreitung der Tropfen einer Flüssigkeit auf der Oberfläche einer anderen. Ann. Physik (Poggendorff), 3, 337–354

Marcus, Y. (1991) LSER Correlation and Prediction of the Distribution of Organic Solutes Between Water and Immiscible Organic Solvents. J. Phys. Chem., 95, 8886–8891

Marr, R., Bouvier, A., Bart, H.-J. (1981) Selective Metal Enrichment by Liquid Membrane Permeation. Ger. Chem. Eng., 4, 209–214

Marr, R., Bart, H.-J. (1982) Metallsalzextraktion. Chem.-Ing.-Techn., 54, 119–129

Marr, R., Draxler, J. (1992) Applications. In: Membrane Handbook, W.S.W. Ho, K.K. Sirkar (Eds), Chap. 39, 701–717, Van Nostrand Reinhold, New York

Marsh, K., Kohler, F. (1985) Thermodynamic Properties of Associated Solutions. J. Mol. Liq., 30, 13–55

Masson, D.O. (1929) Solute Molecular Volumes in Relation to Solvation on Ionization. Phil. Mag., 8, 218–235

Matsumoto, M., Tsuntsuni, Y., Kondo, K., Nakashio, F. (1990) Copper Extraction with a Chelating Reagent in a Hollow Fiber Membrane Extractor. J. Chem. Eng. Japan, 23, 233

Matsumura, M., Märkl, H. (1986) Elimination of Ethanol Inhibition by Pertraction. Biotechnol. Bioeng., 28, 534–541

Matsumura, M. (1991) Pertraction. In: B. Mattiasson, O. Holst (Eds), Extractive Bioconversions, 919, Marcel Dekker, New York

Maurer, G. (1983) Electrolyte Solutions. Fluid Phase Equilibria, 13, 269

Maurer, G. (1996) Phase Equilibria in Chemically Reactive Fluid Mixtures. Fluid Phase Equilibria, 116, 39–51

Maxwell, J.C. (1866) On the Dynamic Theory of Gases. Phil. Trans. Roy. Soc., 157, 49–88

Maxwell, J.C. (1952) The Scientific Papers of James Clerk Maxwell. W.D. Niven (Ed), Dover, New York

Mewes, D., Kunkel, W. (1978) Estimation of the Height of the Dispersed Phase Under the Plates of a Sieve-Plate Extraction Column. Ger. Chem. Eng., 2, 111–115

Mewes, D. Pilhofer, T. (1978) Vorausberechnung der fluiddynamischen Eigenschaften von Siebbodenextraktionskolonnen ohne Pulsation. Chem.-Ing.-Techn., 50, 203–211

Mewes, D., Pilhofer, T. (1979) Prediction of Fluid-Dynamic Properties of Unpulsed Sieve-plate Extraction Columns. Ger. Chem. Eng., 2, 69–76

Meyer, P., Maurer, G. (1993) Correlation of Partition Coefficients of Organic Solutes Between Water and an Organic Solvent. An Application of the LSER. Ind. Eng. Chem. Res., 32, 113–131

Meyer, P., Maurer, G. (1995) Correlation and Prediction of Partition Coefficients of Organic Solutes Between Water and an Organic Solvent with a Generalized Form of the LSER. Ind. Eng. Chem. Res., 34, 373–381

Milero, F.J. (1972) The Partial Molar Volumes of Electrolytes in Aqueous Solutions. In: Water and Aqueous Solutions, R.A. Horne (Ed), Wiley-Interscience, New York

Miller, G., Readett, D., Hutchinson, P. (1996) Entrainment Coalescing in Copper SX Circuits. In: Value Adding Through SX,. D.C. Shallcross et al. (Eds), 795 - 800, The University of Melbourne, Melbourne

Mock, B., Evans, L.B., Chen, C.C. (1986) Thermodynamic Representation of Phase Equilibria of Mixed Solvent Electrolyte Systems. AIChE J., 32, 10, 1655–1664

Modes, G. (2000) Grundsätzliche Studie zur Populationsdynamik einer Extraktionskolonne auf Basis von Einzeltropfenuntersuchungen. Shaker Verlag, Aachen

Mörters, M. (2000) Zum Stoffübergang in Tropfen bei der Reaktivextraktion. Shaker Verlag, Aachen

Mörters, M., Bart, H.-J. (2000) Extraction Equilibria of Zinc with Bis(2-ethylhexyl)phosphoric acid. J. Chem. Engng. Data, 45, 1, 82–85

Mörters, M., Bart, H.-J. (2000a) Examination of the Multi Component Mass Transfer in Reactive Heavy Metal Extraction. In: Transportmechanism across Fluid Interphases, DECHEMA Monographs Vol. 136, Wiley-VCH, Weinheim, 373–386

Mörters, M., Bart, H.-J. (2000b) Fluorescence-Indicated Mass Transfer in Reactive Extraction. Chem. Eng. Technol., 23, 4

Moosbrugger, T. (1991) Kritischer Vergleich literaturbekannter Modelle für die Extraktionskinetik des Systems Zink, Di(2-ethyl-hexyl)phosphorsäure, n-Dodecan. Master thesis, Technical University Graz, Graz

Murthy, C.V.R., Perez de Ortiz, E.S. (1986) Comparison Between Single Drops and Stirred Cell Technique in the Modeling of the Stripping of Zinc from Di(2Etylhexyl)Phosphoric Acid. Proc. ISEC '86, Munich, Vol. II, 353–360, DECHEMA (Ed), Frankfurt

Nagata, I., Miyamoto, K. (1993) VLE and Excess Enthalpies in 1-Alcohol and n-Alkane Mixture. Fluid Phase Equilibria, 89, 173–186

Nakanishi, K. (1978) Prediction of Diffusion Coefficients of Nonelectrolytes in Dilute Solution Based on Generalized Hammond–Stokes Plot. Ind. Eng. Chem. Fundam., 17, 253–256

Naser, S.F., Fournier, R.L. (1988) A Numerical Evaluation of a Hollow Fiber Extractive Fermenter Process for the Production of Ethanol. Biotechnol. Bioeng., 32, 628–638

Neumann, R.D., Park, S.J. (1992) Characterization of Association Microstructures in Hydrometallurgical Nickel Extraction by Di(2-ethylhexyl)phosphoric acid. J. Colloid Interface Sci., 152, 41–52

Neumann, R.D., Park, S.J., Zhou, N.F., Shah, P. (1993) Interfacial Phenomena in Hydrometallurgical SX Systems. In: Solvent Extraction in the Process Industries, D.H. Logsdail, M.J. Slater (Eds), 1689–1696, Elsevier, London

Newman, A.B. (1931a) The Drying of Porous Solids: Diffusion Calculations. Trans. I. Ch. E., 27, 310

Newman, A.B. (1931b) The Drying of Porous Solids: Diffusion and Surface Emission Equations. AIChE J., 27, 203–220

Newman, J. (1991) Electrochemical Systems, 2nd Ed., Prentice-Hall, New Jersey

Nitsch, W., Schoor, van, A. (1983) The Kinetics of Coextraction in the System Uranyl nitrate Nitric acid Tributylphosphate. Chem. Eng. Sci., 31, 12, 1947–1957

Nitsch, W. (1989) Transportprozesse und chemische Reaktionen an fluiden Phasengrenzen. Dechema-Monographie, Bd. 114, 285–302, VCH, Weinheim

Nitsch, W., Plucinski, P., Ehlenspiel, J. (1996) The Mass Transfer of Water between Water-in-Oil Microemulsion and Conjugate Aqueous Phase – a Key to the Mechanism of Reverse Micellar Extraction. In: Value Adding Through Solvent Extraction, D.E. Shallcross, R. Paimin, L.M. Prvcic (Eds), Melbourne, 201–206

Olney, R.B. (1964) Doplet Characteristics in a Countercurrent Extraction. AIChE J., 10, 827

Ortiz, de, E.S.P. (1992) Marangoni Phenomena. In: Science and Practice of Liquid–Liquid Extraction, J.D. Thornton (Ed), Clarendon Press, Oxford, 157–209

Ortner, A., Kronberger, T., Zulehner, W., Bart, H.-J. (1995) Tropfen-Populanz-Bilanzmodell am Beispiel einer gerührten Extraktionskolonne. Chem.-Ing.-Techn., 67, 984–988

Overbeck, J.Th.G., Verhoeckx, G.J., Bruyn, P.L., Lekkerkerker, H.N.W. (1987) On Understanding Microemulsions. Colloid and Interface Sci., 119, 422–441

Park, S.J., Neuman, R.D. (1992) Extraction of Nickel From Nitrate Media with Di(2-ethylhexyl)phoshoric acid Under High Loading Conditions. Sovent Extraction '90 , T. Sekine (Ed), 201–206, Elsevier, Amsterdam

Peng, D.Y., Robinson, P.B. (1976) A new Two-Constant Equation of State. Ind. Eng. Chem. Fundam., 15, 59–64

Pertler, M., Blass, E., Stevens, G.W. (1996) Fickian Diffusion in Binary Mixtures That Form Two Liquid Phases. AIChE J., 42, 4, 910–920

Pfennig, A., Schwerin, A. (1998) Influence of Electrolytes on Liquid–Liquid Extraction. Ind. Eng. Chem. Res., 37, 3180–3188

Pfleiderer, P. (1898) Deutsches Patent Nr 113946

Philipp, A., Hartung, P., Hahn, M. (1997) Wasserbestimmung in Ölen. Labor-Praxis, April, 54–58

Pilhofer, T., Goedl, R. (1977) Belastungsdiagramm für Siebboden-extraktionskolonnen. VT-Verfahrenstechnik, 11, 667–669

Pilhofer, T., Mewes, D. (1979) Siebbodenextraktionskolonnen. VCH, Weinheim

Pitzer, K.S. (1973) Thermodynamics of Electrolytes I. Theoretical Basis and General Equations. J. Phys. Chem., 77, 268

Pitzer, K.S., Mayorga, G. (1973) Thermodynamics of Electrolytes II. Activity and Osmotic Coefficients for Strong Electrolyte with One or Both Ions Univalent. J. Phys. Chem., 77, 19, 2300–2308

Pitzer, K.S. (1975) Thermodynamics of Electrolytes V. Effects of Higher Order Electrostatic Terms. J. Sol. Chem., 4, 249–265

Pitzer, K.S., Roy, R.N., Silvester, L.F. (1977) Thermodynamics of Electrolytes 7. Sulfuric Acid. J. Am. Chem. Soc., 97, 4930

Pitzer, K.S. (1979) In: Activity Coefficients in Electrolyte Solutions. R.M. Pytkowicz (Ed), Vol. I, Chap. 7, 157–208, CRC Press, Florida

Polka, H.-J., Li, J., Gmehling, J. (1994) A g^E Model for Single and Mixed Solvent Electrolyte Systems, Part II. Fluid Phase Equilibria, 94, 115–127

Power, K.L. (1971) Operation of the First Liquid Ion-Exchange and Electrowinning Plant. In: Proc. ISEC '71, Vol. II, 1409–1415, Soc. Chem. Ind. (Ed), London

Prasad, R., Sirkar, K.K. (1988) Dispersion-Free Solvent Extraction with Microporous Hollow-Fiber Modules. AIChE J., 34, (2), 177–188

Prasad, R., Sirkar, K.K. (1989) Hollow Fiber Solvent Extraction of Pharmaceutical Products: A Case Study. J. Membr. Sci., 47, 235–259

Prasad, R., Sirkar, K.K. (1992) Membrane-Based Solvent Extraction. In: W.S.W. Ho, K.K. Sirkar (Eds), 727–763, Membrane Handbook, Chapman & Hall, New York

Prausnitz, J.M. (1969) Molecular Thermodynamics of Fluid-Phase Equilibria. Prentice Hall, New Jersey

Prigogine, I., Defay, R. (1954) Chemical Thermodynamics. Longmans/Green, London

Proc. ISEC '71 (1971) The Hague, Ed. Society of Chemical Industry, London

Proc. ISEC '74 (1974) Lyon, Ed. Society of Chemical Industry, London

Proc. ISEC '77 (1979) Toronto, CIM Special Volume 21, Can. Inst. Min. Met., Montreal
Proc. ISEC '80 (1980) Liege, Ed. Ass. des Ingenieurs Sortis de l'Universite de Liege, Liege
Proc. ISEC '83 (1983) Denver, Ed. AIChE, New York
Proc. ISEC '86 (1986) Munich, Ed. Dechema, Frankfurt
Proc. ISEC '88 (1988) Moscow, Ed. Nauka, Moscow
Proc. ISEC '90 (1992) Kyoto, Solvent Extraction 1990, T. Sekine (Ed), Elsevier, Amsterdam
Proc. ISEC '93 (1993) York, Solvent Extraction in the Process Industries. D.H. Logsdail & M.J. Slater (Eds), New York, Society of Chemical Industry, Elsevier, London
Proc. ISEC '96 (1996) Melbourne, Value Adding Through Solvent Extraction. University of Melbourne, D.E. Shallcross, R. Paimin, L.M. Prvcic (Eds), Melbourne
Proc. ISEC '99 (2000) Barcelona, Solvent Extraction for the 21st Century, Elsevier, Amsterdam (in press)
Quayle, O.R. (1953) The Parachors of Organic Compounds: An Interpretation and Catalogue. Chem. Rev., 53 , 439–589
Reed, B.W., Klassen, R., Jansen, A.E., Akkerhuis, J.J., Bult, B.A., Oesterholt, F.I.H.M. (1994) Removal of Hydrocarbons from Waste Water by Membrane Extraction. The AIChE Spring National Meeting 1994, Atlanta
Reed, B.W., Semmens, M.J., Cussler, E.L. (1995) Membrane Contactors. In: R.D. Noble, S.A. Stern (Eds), Membrane Separations Technology Principles and Applications, 474, Elsevier, Amsterdam
Reid, R.C., Prausnitz, J.M., Poling, B.E. (1988) The Properties of Gases & Liquids, International Edition; 4th Ed., McGraw-Hill Inc., Singapore
Renon, H., Prausnitz, J.M. (1968) Local Compositions in Thermodynamic Excess Functions for Liquid Mixtures. AICHE J., 14, 135
Ridgeway, K., Thorpe, E.E. (1983) Use of Solvent Extraction in Pharmaceutical Manufacturing Processes. In: Handbook of Solvent Extraction, T.C. Lo, M.H.I. Baird, C. Hanson (Eds), J. Wiley & Sons, New York, 583–591
Ritcey, G.M., Lucas, B.H. (1974) Diluents and Modifiers – Their Effect on Mass Transfer and Separation. In: Proc. ISEC '74, Vol. III, 2437–2481, Soc. Chem. Ind., London
Ritcey, G.M., Ashbrook, A.W. (1979) Solvent Extraction Principles and Applications to Process Metallurgy, Vol. I & II, Elsevier, New York
Ritcey, G.M. (1998) Private Communication
Robbins, L.A., Cusack, R.W. (1997) Liquid–Liquid Extraction Operations and Equipment. In: Perry's Chemical Engineering Handbook, R.H. Perry, D.W. Green (Eds), Chap. 15, McGraw Hill, New York
Robinson, R.A., Stokes, R.M. (1965) Electrolyte Solutions. Butterworths, London
Rod, V. (1966) Calculating Mass Transfer with Longitudinal Mixing. Brit. Chem. Eng. (2), 6, 483–488
Rogers, R.D., Eiteman, M.A. (Eds) (1995) Aqueous Biphasic Separations: Biomolecules to Metal Ions, Plenum Press, New York
Rogers, R.D., Bauer, C.B., Bond, A.H. (1995) Novel Polyethylene Glucol-Based Aqueous Biphasic Systems for the Extraction of Strontium and Cesium. Sep. Sci. Technol., 30, 1203
Rogers, R.D., Bond, A.H., Griffin, S.T., Horwith, E.P. (1996) New Technologies for Metal Ion Separations: Aqueous Biphasic Extraction Chromatography, Part I: Uptake of Pertechnate. Solv. Extr. Ion Exchange, 14, 919

Roos, M. (2000) Zum Einfluss von Salzen auf die Reaktivextraktion organischer Säuren. Shaker Verlag, Aachen

Roos, M., Bart, H.-J. (2000a) Extraction of Acetic Acid with Tri-n-octylamine – Physical Properties and Phase Equilibria, J. Chem. Engng. Data, submitted

Roos, M., Bart, H.-J. (2000b) Das LAP-Modell – Ein allgemein konsistenter Ansatz zur spezifischen Beschreibung von Grenzflächenkonzentrationen bei der Reaktivextraktion. Fachausschuss "Thermische Zerlegung von Gas- und Flüssigkeitsgemischen", 5./8.4.2000, Wernigerode

Ruff, K. (1974) Grenze zwischen Blasengasen und Strahlgasen bei niedrigviskosen Flüssigkeiten und konstantem Gasvolumen-Durchsatz. Chem.-Ing.-Techn., 46, 769–771

Rutten, P.W.M. (1992) Diffusion in Liquids. PhD thesis, Delft University, Delft

Rydberg, J., Musikas, C., Choppin, G.R. (Eds) (1992) Principles and Practices in Solvent Extraction. Marcel Dekker, New York

Sainz-Diaz, C.I., Galvez-Ruano, E., Hernandez-Laguna, A., Bellanato, J. (1995) Synthesis Molecular Structure and Spectroscopical Properties of Alkanephosphoric Derivates. J. Org. Chem., 60, 74–83

Sainz-Diaz, C.I., Klocker, H., Bart, H.-J., Marr, R. (1996) New Approach in the Modelling of the Extraction Equilibria of Zinc with Bis(2-ethylhexyl) phosphoric acid. Hydrometallurgy, 42, 1–11

Sato, T., Nakamura, T. (1972) The Complexes Formed in the Divalent Metal–Sulfuric acid-Di-(2-ethylhexyl)phosphoric acid Extraction Systems Cobalt(II), Nickel(II) and Copper(II) Complexes. J. Inorg. Nucl. Chem., 34, 3721–3730

Sato, Y., Akiyoshi, Y., Kondo, K., Nakashio, F. (1990) A Novel Membrane Extractor using Hollow Fibers for Separation and Enrichment of Metals. J. Chem. Eng. Japan, 23, 23–29

Scamehorn, J.F., Harwell, J.H. (1988) Surfactant-Based Treatment of Aqueous Process Streams. In: Surfactants in Chemical/Process Engineering, D.T. Wasan, M.E. Ginn, D.O. Sah (Eds), Surfactant Science Series, Vol. 28, 77–125, Marcel Dekker, New York

Schick, M.J., Fowkes, F.M. (Eds) (1966–1999) Surfactant Science Series, Marcel Dekker, Vols. 1–83, New York

Schöneberger, A., Bart, H.-J., Nanoti, A., Wildberger, A. (2000) Extraction Chromatography: Equilibria and Kinetics. In: ISEC '99 – Solvent Extraction for the 21st Century, Elsevier, in press

Schöner, P., Plucinski, P., Nitsch, W., Daiminger, U. (1998) Mass Transfer in the Shell Side of Cross Flow Hollow Fiber Modules. Chem. Eng. Sci., 53, 13, 2319–2326

Schröter, J., Bäcker, W., Hampe, M. (1998) Stoffaustauschmessungen an Einzeltropfen und an Tropfenschwärmen in einer Gegenstrommesszelle. Chem.-Ing.-Techn., 70, 279–283

Schwuger, M.J. (1996) Lehrbuch der Grenzflächenchemie. G. Thieme Verlag, Stuttgart

Scott, R.L. (1956) Corresponding States Treatment of Nonelectrolyte Solutions. J. Chem. Phys., 25, 193–205

Seibert, A.F., Py, X., Mshewa, M., Fair, J.R. (1993) Hydraulics and Mass Transfer Efficiency of a Commercial-Scale Membrane Extractor. Sep. Sci. Technol., 28, (1–3), 343–359

Seibert, A.F., Fair, J.R. (1997) Scale-Up of Hollow Fiber Extractors. Sep. Sci.Technol., 32 (1–4), 573–583

Sengupta, A., Reed, B.W., Seibert, F. (1994) Liquid–Liquid Extraction Studies on Semi-Commercial Scale Using Recently Commercialized Large Membrane Contactors and Systems. The AIChE Annual Meeting 1994, San Francisco

Shelley, F., Quan, C. (1992) Performance of Two New Nil Aromatic Diluents in the Recovery of Copper, Uranium and Rare Earths. In: Solvent Extraction 1990, T. Sekine (Ed), 381–386, Elsevier, Amsterdam

Shukla, R., Kang, W., Sirkar, K.K. (1989) Acetone-Butanol-Ethanol (ABE) Production in a Novel Hollow Fiber Fermentor–Extractor. Biotechnol. Bioeng., 34, 1158–1166

Sirkar, K.K. (1997) Membrane Separation Technologies: Current Developments. Chem. Eng. Commun., 157, 145–18400

Skelland, A.H.P., Conger, W.L. (1973) A Rate Approach to Design of Perforated Plate Extraction Columns. Ind. Eng. Chem. Proc. Des. Dev., 12, 4, 448–454

Slater, M.J. (1994) Rate Coefficients in Liquid–Liquid Extraction Systems. In: Liquid Extraction Equipment, J.C. Godfrey and M.J. Slater (Eds), 45–94, J. Wiley & Sons, Chichester

Slater, M.J. (1995) A Combined Model of Mass Transfer Coefficients for Contaminated Drop Liquid–Liquid Systems. Can. J. Chem. Eng., 73, 462–469

Smelov, V.S., Lanin, V.P. (1969) Dissociation of Bis(2-ethylhexyl) hydrogen phosphate in an Aqueous Solution. Radiokhimiya, 11 (4), 445–447

Soave, G. (1972) Equilibrium Constants from a Modified Redlich–Kwong Equation of State. Chem. Eng. Sci., 27, 1197–1203

Soler, J., Urtiaga, A.M., Irabien, A., Ortiz, I. (1996) Hollow Fiber Non-Dispersive Solvent Extraction for the Recovery of Nickel and Cadmium. In: Value Adding Through Solvent Extraction, D.C. Shallcross, R. Paimin, L.M. Prvcic (Eds), Vol. II, 851–856, The University of Melbourne, Melbourne

Sorensen, J.M., Magnussen, T., Rasmussen, P., Fredenslund, A. (1979) Liquid–Liquid Equilibria Data: Their Retrieval, Correlation and Prediction, Part II: Correlation. Fluid Phase Equilibria 3, 47–82

Sorensen, J.M., Arlt, W. (1979–1980) Liquid–Liquid Equilibrium Data Collection. Chem. Data Series, DECHEMA, Frankfurt

Stefan, J. (1871) Über das Gleichgewicht und die Bewegung, insbesondere die Diffusion von Gasmengen. Sitzungsberichte Akad. d.Wiss., Wien, 63–124

Steiner, L. (1986) Mass Transfer Rates from Single Drops and Drop Swarms. Chem. Eng. Sci., 41, 8, 1979–1986

Sterling, C.V., Scriven, L.E. (1959) Interfacial Turbulence: Hydrodynamic Instability and Marangoni Effect. AIChE J., 5, 104–106

Steward, W.E., Prober, R. (1964) Matrix Calculation of Multicomponent Mass Transfer in Isothermal Systems. Ind. Eng. Chem. Fund., 3, 224–235

Stokes, R.H. (1977) Interpretation of the Thermodynamic Spectroscopic and Dielectric Properties in Solutions of Ethanol in Cyclohexane in Terms of Association. J. Chem. Thermodyn., 8, 1140–1148

Strikovsky, A.G., Jerabeth, K., Cortina, J.L., Saestre, A.M., Warshawsky, A. (1996) Solvent Impregnated Resins Containing Dialkyl dithio-phosphoric acid on Amberlite XAD-2: Extraction of Copper and Comparison to Liquid–Liquid Extraction. Reactive & Functional Polymers, 28 (2), 149–158

Sunderrajan, S., Hall, C.K., Freeman, B.D. (1996) Estimation of Mutual Diffusion Coefficients in Polymer/Penetrant Systems Using Nonequilibrium Molecular Dynamics Simulations. J. Chem. Phys., 105 (4), 1621–1632

Tavlarides, L.L., Nandkumar, V.D. (1997) Zinc, Cadmium and Lead Separation from Aqueous Streams Using Solid-Phase Extractants. Ind. Eng. Chem. Res., 36, 399–406

Taylor, R., Webb, D.R. (1980) On the Relationship Between the Exact and Linearized Solutions of the Maxwell–Stefan Equations for the Multiccomponent Film Model. Chem. Eng. Commun., 7, 287–299

Taylor, R., Kooijman, H. (1991) Some Issues Concerning Diffusion in Multicomponent Systems. In: Phase-Interface Phenomena in Multiphase Flow, G.F. Hewitt, F. Mayinger, J.R. Riznic (Eds), Hemisphere, New York, 495–514

Taylor, R., Krishna, R. (1993) Multicomponent Mass Transfer. J. Wiley & Sons, New York

Thiel, P. (1995) Diffusionsmessungen in Flüssigkeiten mittels Taylor–Dispersionsmethode. PhD thesis, Martin-Luther-University Halle-Wittenberg, Halle

Thomas, E.R., Eckert, C.A. (1984) Prediction of Limiting Activity Coefficients by the Modified Separation of Cohesive Energy Density Model and UNIFAC. Ind. Eng. Chem. Proc. Des. Dev., 23, 194–209

Thomson, W. (Lord Kelvin) (1871) The Influence of Wind on Waves in Water Supposed Frictionless. Phil. Mag. 10, 330–333

Thornton, J.D. (1956) Spray Liquid–Liquid Extraction Columns. Prediction of Limiting Hold-up and Flooding Rates. Chem. Eng. Sci., 5, 201

Thornton, J.D. (Ed) (1992) Science and Practice of Liquid–Liquid Extraction, Vol. I&II, Oxford University Press, Oxford

Tian, H.S., Wang, X.T., Su, Y.F. (1992) Influence of Diluents on the Extraction of Fluorine from Wet Phosphoric Acid with Trioctylamine. In: Solvent Extraction 1990, T. Sekine (Ed), 387–392, Elsevier, Amsterdam

Toor, H.L. (1964) Solution of the Linearized Equations of Multicomponent Mass Transfer. AIChE J., 10, 448–455 & 460–465

Traving, M., Bart, H.-J. (2000) Recovery of Organic Substances from Fermentation Broths Using Reactive Sorption. In: Solvent Extraction for the 21st Century, Elsevier, in press

Traving, M. (2000) Reaktivsorption: Ein Verfahren zur Aufarbeitung von Fermentationen und hochverdünnten Lösungen. Shaker Verlag, Aachen

Treybal, R.E. (1963) Liquid Extraction. McGraw Hill, New York

Tyn, M.T., Calus, W.F. (1975) Diffusion Coefficients in Dilute Binary Liquid Mixtures. J. Chem. Eng. Data, 20, 106–109

Tyrrell, H.J.V., Harris, R.Q. (1984) Diffusion in Liquids. Butterworths, New York

Ul'yanov, V.S., Sviridova, R.A. (1963) Dissociation, Dimerization and Distribution of Bis(2ethyl-hexyl)phosphoric acid in the System Octane–0.1 M $NaClO_4$ Solution–$HClO_4$. Radiokhimiya, 5, 419

Valentas, K.J., Amundson, N.R. (1966) Breakage and Coalescence in Dispersed Phase Systems. Ind. Eng. Chem. Fundam., 5, 533–542

Van Laar, J.J. (1910) The Vapor Pressure of Binary Mixtures. Z. Phys. Chem. 72, 723–751

Van Ness, H.C., Abbott, M.M. (1997) Thermodynamics. In: Perry's Chem. Engineering Handbook, 7th Ed., R.H. Perry, D.W. Green (Eds), Section 4, McGraw Hill, New York

Van Oss, C.J., Chandhury, M.K., Good, R.J. (1989) The Mechanism of Phase Separation of Polymers in Organic Media–Apolar and Polar Systems. Sep. Sci. Technol., 24, 15

Vandegrift, G.F., Horwitz, E.P. (1977) The Mechanism of Interfacial Mass Transfer of Calcium in the System: Di-(2-ethylhexyl)phosphoric acid in Dodecane-Dilute Nitric acid. J. Inorg. Nucl. Chem., 39, 1425–1432

Vatai, G., Tekic, M.N. (1991) Membrane-Based Ethanol Extraction with Hollow-Fiber Module. Sep. Sci. Technol., 26 (7), 1005–1011

Veglio, E., Slater, M.J. (1996) Design of Liquid-Liquid Extraction Columns for the Possible Test System Zn/D2EHPA in n-Dodecane. Hydrometallurgy, 42, 177–195

Ven-Lucassen, van de, I.M.J.J. (1999) Diffusion Coefficients in Liquid–Systems. PhD thesis, TU Eindhoven, Universiteits Drukkerij, Eindhoven

Vignes, A. (1965) Hydrodynamique des dispersions – mouvement d´un globule dans un fluide immobile et infini. Génie Chimique, 93, 129–142

Vignes, A. (1966) Diffusion in Binary Solutions. Ind. Eng. Chem. Fundam., 5, 189–199

Villaescusa, I., Salvadó, V., de Pablo, J. (1996) Liquid–Liquid and Solid–Liquid Extraction of Gold by Trioctylammonium chloride (TOMACl) Dissolved in Toluene and Impregnated on Amberlite XAD-2 Resin. Hydrometallurgy, 41, 303–311

Von Reden, C. (1998) Multicomponent Mass Transfer in Liquid Extraction. Shaker Verlag, Aachen

Wachter, B., Bart, H.-J., Moosbrugger, T., Marr, R. (1993) Reactive Liquid–Liquid Test System Zn/Di(2-ethylhexyl) phosphoric acid/n-Dodecane: Equilibrium and Kinetics. Chem. Eng. Technol., 16, 413–421

Wachter, B. (1996) A Possible Recommended Reactive Liquid–Liquid System Zinc/ Di-(2–ethylhexyl)phosphoric acid/n-Dodecane: Modelling of Equilibrium by Means of SIMUSOLV. PhD thesis, Graz University, Graz

Walter, H., Johansson, G. (Eds) (1994) Aqueous Two-Phase Systems. In: Methods in Enzymology, Vol. 228, Academic Press, San Diego

Wang, K.L., Cussler, E.L. (1993) Baffled Membrane Modules Made with Hollow Fiber Fabric. J. Membr. Sci., 85, 265–278

Warshawsky, A., Patchronik, A. (1978) Impregnated Resins: Metal Ion Complexing Agents Incorporated Physically in Polymeric Matrices. Israel J. Chemistry, 17, 307–315

Warshawsky, A. (1979) Solvent Impregnated Resins: Bridging the Gap Between Liquid and Solid Extraction. ISEC '77, CIM Spec., Vol. 21, Can. Inst. Min. Met., Toronto, 127–130

Warshawsky, A., Berkowitz, H. (1979) Hydroxime Solvent-Impregnated Resins for Selective Copper Extraction. Metallurgy, C36–C43

Way, D.J., Noble, D.R. (1992) Facilitated Transport. In: Membrane Handbook, W.S.W. Ho, K.K. Sirkar (Eds), Chap. 44, 833–865, Van Nostrand Reinhold, New York

Weatherly, L.R. (1992) Some Current Developments and Extraction Techniques. In: Science and Practice of Liquid–Liquid Extraction, J.D. Thornton (Ed), Vol. 2, 353–422, Clarkson Press, Oxford

Weidlich, U., Gmehling, J. (1987) A Modified UNIFAC Model, Part I, Prediction of VLE, h_E and γ^∞. Ind. Eng. Chem. Res., 26, 1372

Wedler, G. (1987) Lehrbuch der physikalischen Chemie. 3. Aufl., VCH Verlagsgesellschaft mbH, Weinheim

Wesselingh, J.A., Bollen, A.M. (1999) Single Particles, Bubbles and Drops: Their Velocities and Mass Transfer Coefficients. Trans. Inst. Chem. Eng., 77, Part A, 89–96

Wesselingh, J.A., Krishna, R. (2000) Mass Transfer in Multicomponent Mixtures. Delft University Press, Delft

Wickramasinghe, S.R., Semmens, M.J., Cussler, E.R. (1992) Mass Transfer in Various Hollow Fiber Geometries. J. Membr. Sci., 69, 235–250

Wilke, C.R., Chang, P. (1955) Correlation of Diffusion Coefficients in Dilute Solutions. AIChE J., 1, 264–270

Wisniak, J., Tamir, A. (1980) Liquid–Liquid Equilibrium and Extraction. Phys. Sci. Data 7, Elsevier, New York

Yang, M.C., Cussler, E.L. (1986) Designing of Hollow-Fiber Contactors. AIChE J., 32, 1910–1915

Yang, Z.-F., Guha, A.K., Sirkar, K.K. (1996) Novel Membrane-Based Synergistic Metal Extraction and Recovery Processes. Ind. Eng. Chem. Res., 35, 1383–1394

Ye, M.H. (1988) PhD thesis, Tsinghua University, Beijing

Ye, M.H., Fei, W.Y., Dai,Y.Y., Wang, J.D. (1996) A Group Contribution Method for the Prediction of Diffusion Coefficients in Liquids. In: Value Adding Through Solvent Extraction, Vol. 1, 69–74, University of Melbourne, D.C. Shallcross, R. Paimin, L.M. Prvicic (Eds), Melbourne

Yoshizuka, K., Kondo, K., Nakashio, F. (1986) Effect of Interfacial Reaction on Rates of Extraction and Stripping in Membrane Extractor Using a Hollow Fiber. J. Chem. Eng. Japan, 19 (4), 312–318

Yoshizuka, K., Yasukawa, R., Koba, M., Inoue, K. (1995) Diffusion Model Accompanied with Aqueous Homogeneous Reaction in Hollow Fiber Membrane Contractor. J. Chem. Eng. Japan, 28 (1), 59–65

Young, T.F., Smith, M.B. (1954) Thermodynamic Properties of Mixtures of Electrolytes in Aqueous Solutions. J. Phys. Chem., 58 (9), 714–724

Yun, C.H., Prasad, R., Sirkar, K.K. (1989) Solvent Extraction of Priority Organic Pollutants Using Hollow Fiber Membranes. AIChE National Meeting 1989, Philadelphia

Yun, C.H., Prasad, R., Guha, A.K., Sirkar, K.K. (1993) Hollow Fiber Solvent Extraction Removal of Toxic Heavy Metals from Aqueous Waste Water Streams. Ind. Eng. Chem. Res., 1186–1195

Zermaitis, J.F., Clark, D.M., Marshall, R., Scrivner, N.C. (1986) Handbook of Aqueous Electrolyte Thermodynamics. AIChE, New York

Ziegenfuß, H., Maurer, G. (1994) Distribution of Acetic Acid Between Water and Organic Solutions of TOA. Fluid Phase Equilibria, 102, 211–255

Author Index

Subject Index